# THE NEWTON PAPERS

23

To make excellent Ink.

Rx ½ lb of Galls cut in pieces or grossly beaten, ¼ lb of Gumm Arabick cut or broken. Put 'em into a Quart of strong bear or Ale. Let 'em stand a month stopt up, stirring them now & then. At ye end of the month put in ℥1 or ℥1½ of copperas (Too much copperas makes ye ink apt to turn yellow.) Stir it & use it. Stop it up for some time with a paper prickt full of holes & let it stand in ye sunn. When you take out ink put in so much strong bear & it will endure many years. Water makes it apt to mold. Wine does not. The air also if it stand open inclines it to mold. With this Ink new made I wrote this.

A recipe for "excellent ink" in Newton's hand. MS Add. 3975 f23. Reproduced by kind permission of the Syndics of Cambridge University Library.

# THE NEWTON PAPERS

## The Strange and True Odyssey of Isaac Newton's Manuscripts

Sarah Dry

OXFORD
UNIVERSITY PRESS

OXFORD
UNIVERSITY PRESS

Oxford University Press is a department of the
University of Oxford. It furthers the University's objective of
excellence in research, scholarship, and education
by publishing worldwide.

Oxford New York
Auckland Cape Town Dar es Salaam Hong Kong Karachi
Kuala Lumpur Madrid Melbourne Mexico City Nairobi
New Delhi Shanghai Taipei Toronto

With offices in
Argentina Austria Brazil Chile Czech Republic France Greece
Guatemala Hungary Italy Japan Poland Portugal Singapore
South Korea Switzerland Thailand Turkey Ukraine Vietnam



Published in the United States of America by
Oxford University Press
198 Madison Avenue, New York, NY 10016



Library of Congress Cataloging-in-Publication Data
Dry, Sarah, 1974–, author.
The Newton papers : the strange and true odyssey
of Isaac Newton's manuscripts / Sarah Dry.
pages cm
Includes bibliographical references and index.
ISBN 978-0-19-995104-8 (hardback)
1. Newton, Isaac, 1642–1727—Manuscripts.
2. Newton, Isaac, 1642–1727—Psychology.
3. Science—History. I. Title.
QC16.N7D79 2014
530.092—dc23 2013039930

3 5 7 9 8 6 4 2

Printed in the United States of America
on acid-free paper

FOR MY PARENTS

# Contents

# Acknowledgments

In tracing the history of Newton's manuscripts, I have relied on the assistance of librarians and archivists from around the world. Much of this book was written in the lucky surrounds of the Huntington Library in San Marino, California, where I owe thanks to Dan Lewis and Jaeda Snow for making things easy for me. Staff at the Sherman Fairchild and Millikan Libraries at the California Institute of Technology were similarly welcoming. For making material from their collections available to me, I would also like to thank Rachel Misrati and Yael Okun at the National Library of Israel; Dee Stonberg at the Horn Library of Babson College; Adam Perkins and Frank Bowles at Cambridge University Library; Ryan Cronin and Kathryn McKee at St. John's College Library, Cambridge; and the staff at the Hampshire Record Office. I am especially grateful to Micah Anshan, who generously shared his research in the Yahuda archive, and to Diana Kormos-Buchwald of the Einstein Papers Project at Caltech for translating much of the Yahuda-Einstein correspondence.

One of the best parts of writing this book has been talking (and emailing) about it with historians like Mordechai Feingold, Scott Mandelbrote, Simon Schaffer, Stephen Snobelen, Larry Stewart, Andy Warwick, and Robert Westman. I am grateful for their input. George Smith read the entire manuscript and made it much better, sharing his personal recollections of I. B. Cohen along the way. Jeremy Norman, Robert Harding, Arnold Hunt, and Paul Quarrie graciously offered their hard-won expertise from the world of book dealing. David Castillejo kindly shared memories of his role in bringing the Newton papers to light. Any errors that remain are, alas, my own.

Kit Ward was my dream agent: astute, enthusiastic, and wonderfully encouraging. I was very lucky to find her. Sadly, she died unexpectedly in November 2012. I am sorry I couldn't have known her longer and grateful to have known her at all.

Tim Bent, my editor at Oxford University Press, saw the shape of the story and helped me to see it too. It is a much better book for his thoughtful interventions.

My family supported me through the calms and the rapids of the research and writing process and always reminded me to keep my eye on the larger story. A litany of Drys—Katie, Cecie, Paul, Rachel, and Bonnie—read and commented on parts of the manuscript. My brother-in-law, Guillermo Bleichmar, went further and read the whole thing, as did David Stern, who is like family. I am very grateful to all of them for taking Newton into their lives.

There are many senses in which this book would not have been written without Rob Iliffe. As editor of the Newton Project, he has been largely responsible for bringing the private Newton out of the dark and into the digital light. As a scholar of Newton's religion, he has taught me the pleasures to be had in the company of a radical Protestant. And as my husband, he has rallied beside me day by day. It was Rob who first told me more than ten years ago the story of Newton's papers and encouraged me to write a book about it. That idea took a while to germinate, but in the meantime, other seeds grew. Jacob Iliffe, born in 2009, knows that Newton was a man who died, which seems a pretty good place to start.

Finally, I want to thank my parents, Cecie and Paul Dry, for the gift of continuous love and support. There is no better inheritance.

## *Note on Sources*

They call it the Newton industry for a good reason. Historical treatments of Isaac Newton and his influence abound, and this book could not have been prepared without them. It would not be practical to reference here all the scholarship on which I have relied. For readers who want to learn more, Richard Westfall's *Never at Rest* (Cambridge University Press, 1980) remains the authoritative (mostly) scientific biography of Newton, while Rob Iliffe's *Newton: A Very Short Introduction* (Oxford University Press, 2007) provides a succinct synopsis of Newton's life and work. In recent years a number of major scholarly works drawing on Newton's unpublished papers have appeared, including Niccolo Guicciardini's *Isaac Newton on Mathematical Certainty and Method* (MIT Press, 2009), Jed Buchwald and Mordechai Feingold's *Newton and the Origin of Civilization* (Princeton University Press, 2012) and Rob Iliffe's *High Priest of Nature: The Heretical Life of Isaac Newton* (Oxford University Press, forthcoming). Finally, today it is also possible to encounter Newton directly online at the Newton Project, the Chymistry of Isaac Newton, and the Cambridge Digital Library. I relied extensively on these remarkable resources in the writing of this book. I recommend taking the opportunity to accompany Newton as he goes about his business. There is no substitute for seeing the papers for oneself.

# THE NEWTON PAPERS

# Prologue: Keynes at the Sale

John Maynard Keynes arrived late in the day on July 13, 1936, to the Sotheby auction on New Bond Street in London. Keynes was a busy man—economic adviser to government; teacher of economics at Cambridge, where he was also the financial administrator of King's College; and self-anointed squire of his country house at Tilton in the Sussex countryside, a gathering place for the Bloomsbury set. Just months earlier he had published *The General Theory of Employment, Interest, and Money*, a book that revolutionized monetary policy and transformed Keynes into the most influential economist of his generation. It was only a last-minute tip from his younger brother, Geoffrey, a surgeon and bibliophile, that had alerted him to the Sotheby sale at all. With rumors of another war growing (in March of that year, in violation of the Treaty of Versailles, the Germans had occupied the Rhineland), he could easily have decided that he had more important things to do than bid on a stack of dusty old papers.

But Keynes was an aesthete who made time for matters of taste as well as state. While Bloomsbury sometimes disparaged him (most notably Virginia Woolf, who commented haughtily on Keynes's habit of eating crumpets with butter, "It is this kind of tallow grease grossness in him that one dislikes"[1]), he was nonetheless one of them. The house at 46 Gordon Square in London—a pole star in the cosmology of Bloomsbury—had been in his name for several years. He had occupied the spacious second-floor bedroom and study since the end of the Great War, though the murals of his former lover Duncan Grant had been whitewashed soon after Keynes's marriage to the Russian ballerina Lydia Lopokova in 1925.

By the time Keynes slipped into the back of the auction room on New Bond Street, the sale was well under way. Keynes bid rather feebly for a few batches of papers. His acquisitive appetite whetted, he returned the next day and stepped up his buying considerably. Though he moved confidently, even aggressively, on some items, he let others go. In all, Keynes acquired thirty-eight lots from the sale. The rest of the papers were sold to professional dealers in whose mutual interest it was to keep bidding sedate and prices low.

On the whole, interest at this sale was lukewarm. This might seem somewhat shocking to us, for it involved more than three hundred lots of Isaac Newton's manuscripts—a trove of virtually unknown private writings by the most celebrated and influential scientist in the world. Containing well over five million words in his own handwriting, these papers had been seen by only a few men since Newton's death in 1727. Altogether they sold for just over £9,000 (£330,000 in today's currency). Today a single sheet of manuscript in Newton's own hand on a scientific subject can bring up to £130,000.

The reason for the lackluster bidding was that few understood what the dusty papers actually contained.

It had not always been so. Years, even centuries earlier, some who knew what was in them had worked hard to conceal their contents, and they had been mostly successful. Aside from brief, troubled glimpses gained by a few men, the papers had remained unexamined, hidden from sight for more than seven generations. The hints they gave were of a Newton altogether stranger than the one entombed in Westminster Abbey as the paragon of English rationality, the man who had simultaneously invented and perfected nothing less than a system for explaining the entire universe. The manuscripts threatened to undermine not just Newton's reputation but, some felt, that of science itself.

This book tells the story of Newton's papers, hidden for centuries, dispersed at auction, and then painstakingly reconstructed. It tracks the history of thoughts that Newton put to paper across the long span of his life (he died at eighty-four), starting nearly three hundred years ago with Newton's death and extending to the present day. Following the papers means following those who were determined to chase the image of Newton through the thickets of his various obsessions. In the process these men, who in addition to Keynes include the inventor of the kaleidoscope, the discoverer of the planet Neptune, the wife of a self-made Yankee business guru, and a Jewish biblical scholar, became obsessed themselves, both with Newton and with the allure and the danger of glorifying a genius, of glorifying any man.

1

# The Death of Newton

Isaac Newton died an old man even by modern standards. By those of his day, when a third to half of all children died before aged sixteen, his was a most conspicuously long life. Born auspiciously on Christmas Day 1642, he survived the ordinary dangers of childhood disease and accident and safely waited out the plague of 1665–66 (which killed some 100,000 people in London alone) in Woolsthorpe, his rural childhood home in the country of Lincolnshire, some one hundred miles north of London. Aside from a period of mental instability during 1693, no serious illness troubled him, leaving him productive and healthy well into his seventies, if less supple of mind than he had been in youth. Only in the final five years of his life did he suffer from physical impediment: a bladder problem that left him incontinent and reliant on a vegetarian diet of broth.

When he felt the end was near, Newton began to prepare. From his ample estate he made specific gifts to his many relatives, including his godson, his second cousin, and the grandson of his uncle. He also burned a number of papers, the mention of which, made by a surviving heir, is casual, as if nothing significant had been destroyed. Given the survival of much rough material, it is hard to imagine what he saw fit to burn that day. In any case, that fact is lost to history, for their contents are unknown.

The last surviving letter that Newton wrote, just over a month before his death, was to the rector of his childhood village, reporting the disappointing news that an assay carried out on a bit of rock from Woolsthorpe was negative for metal; there were no more hidden riches in that village, which had remained close to Newton's heart throughout his life.

Newton had one other, more urgent piece of writing to attend to. Just a few days before his death, a visitor saw him working hard to prepare a clean copy of something for the press. Rather than a mathematical treatise Newton wished to preserve for posterity, the piece in question was a highly detailed chronology of ancient history. It represented the culmination of a life devoted to study. He worked without glasses, in the darkest part of his study. Remarking, "A little light serves me," he discussed his calculations for the timeline with the visitor for nearly an hour.[1]

Newton attended his last meeting of the Royal Society on March 2, 1727. The next morning he received visitors, and that, combined with the excitement of the previous day's meeting, was enough to bring back a violent cough that he had been fighting. His two physicians, the best London had to offer, were called in. They diagnosed the distemper as a stone in the bladder and determined that nothing further could be done to delay the inevitable.

Though his affairs were more or less in order, Newton's death was not peaceful. His biographer and friend William Stukeley described how his pain "rose to such a height that the bed under him, and the very room, shook with his agonys, to the wonder of those that were present." In between these terrible spasms, however, Newton exhibited remarkable poise, smiling and talking to those around him "with his usual Gaiety."

He remained wholly alert until the evening just prior to his death, "as if the Faculties of his Soul were subject to a total Extinction, and could not feel a Decay." One of the final uses to which he put his senses was to refuse the last sacrament, the Eucharist given to a dying person. His whole life was a preparation for another state, he told those at his bedside in explanation for his extraordinary demurral, and he needed no other provision for a journey to another world. One his physicians, Dr. Mead, later reported (to none other than Voltaire) that Newton also confessed on his deathbed that he was a virgin.

When the end finally did come, early in the morning of Monday, March 20, 1727, Newton met it with the "most exemplary, & remarkable patience, truly philosophical, truly christian; & [with] a resignation to the divine Will, equal to his other vertues." Stukeley waxed elegiac that with his death, Newton's spirit had taken flight "through the well known starry orbs" but that his name would live on earth until the sun finally absorbed the planets "in the last conflagration," as Newton himself had predicted.[2]

Meanwhile, and somewhat more hastily, Newton's death set off a local conflagration, as his sundry relatives descended to partake in the spoils

Newton the timeless sage in a terra cotta study for the full-size marble monument by John Rysbrack erected in Westminster Abbey in 1730. © Victoria and Albert Museum, London.

of their famous relative's notable estate. Newton had no full siblings. His closest relative was Catherine Barton, the daughter of his half-sister Hannah, who had moved in with him soon after he came to London and acted as his housekeeper and companion. Ten years before Newton's death, Catherine had married John Conduitt, a man who was to become a pivotal figure in the disposition of Newton's papers.

Newton had left no will. This was not an oversight; his other preparations for his death make it clear that he knew what was coming. Rather he declared that there was no need to leave special instructions for the distribution of his estate. In practice, however, the matter was not simple, and it was made more difficult because the extended Newton clan was as querulous and grasping as any group of would-be inheritors. Their famous relative had died rich.

An inventory of everything that Newton owned was soon prepared. Written on vellum skins sewn together to form a single document, the inventory measures seventeen feet long and five inches wide.[3] It enumerates the contents of Newton's house on St. Martin's Street in London as they were in the gloomy days immediately after his death. In the garret, or attic, were found some mathematical instruments, "Chymical glasses," and a couple of writing desks alongside a jumble of furniture, bed frames and featherbeds, blankets, tables, and chairs. A walnut cabinet and a writing desk with drawers occupied the front room, as well as some two hundred prints, a copper plate, and a small penknife. Dominating the bedroom was what must have been a dramatic "crimson mohair bed compleat with case curtains of crimson Harrateen"—a linen fabric—while in the dining room "eight India back chairs" were presided over by a "figure cut in ivory of Sir Isaac in a glass frame." The great man had left behind forty China plates, six chocolate cups, "one brown teapot," and a kitchen stocked with a "cheese toaster," a frying pan, and a "tin dripping pan." Less mundanely the house contained thirty-nine silver medals, six gold rings, and "twenty six copies of English medals in plaister of Paris," presumably of Newton himself.

In addition to the scientific instruments and the writing desk, evidence for the true nature of the man who had died lay in the books and manuscripts he had left behind, which were duly counted and, as was customary, classified by size. There were a significant number of them: 362 books of the largest size, the folio (roughly 12 x 15 inches); 477 in quarto (half the size); and 1,057 in more intimate octavo, duodecimo, and even smaller formats. The books were valued at £270 (nearly £23,000 today). Also found "together with above" were a mass of papers that Newton had left behind, described solely in the vaguest of terms as "one hundred weight of pamphlets and Wast books." The inventory also noted that there were three "Manuscripts in a box sealed up at the house of John Conduitt Esq." These are "a short chronicle from the first memory of things in Europe" and a "Chronology of the Antients in five chapters" (the manuscript Newton was copying in the days before his death), which were together valued at the not insignificant sum of £250 (more than £21,000 today), along with an unvalued "history of the prophecies in ten chapters... unfinisht." The inventorists added an addendum to this section, noting that there were several boxes in the house containing "many loose papers and Letters" relating, among other things, to Newton's work at the Mint (where he had served as warden and then master for nearly three decades), the manuscripts already mentioned, and Newton's mathematical works. The inventory ended with a flurry of government securities and other investments that Newton held, which totaled the astounding figure of nearly £30,000 (more than £2.5 million today), most of which was in Bank of England stock but a substantial percentage was in South Sea stock and annuities.[4]

It is clear from this inventory just how imposingly books and papers loomed in the intimate surroundings of his home. It is also evident that the manuscripts were, in a very real sense, uncategorizable. However easy it was to parse the more conventional possessions such as furniture and pots and pans that Newton left behind, as well as the valuable investments his descendants clamored for, the manuscripts both demanded to be acknowledged and refused to be readily accounted for. They were, from the beginning, a parenthetical aside, even a nuisance.

Newton most likely intended to leave his papers in the purgatory in which they landed after his death. As mentioned, despite having ample time to prepare one, he left no will. Nor did he destroy his papers. They contained not only his mathematical and scientific notes but writing on theology, alchemy, and Church history, which, if revealed, would have had dramatic repercussions for his reputation and that of his executors. Evidently Newton believed that what they contained was important, too important to be lost, but also too dangerous to publish right away. It was untenable for him to own up to their contents during his lifetime but unconscionable to destroy them, for they contained a form of truth that he had worked long and hard to unearth.

Though it is strange to consider that Newton would have left a lifetime of writing in such an ambiguous state, he had good reason to do so. He understood that his views were not easy to fathom, but this in no way diminished (and indeed may have strengthened) his conviction in their essential truth. As we will see, given their contents, any clear instructions could have caused his executors significant problems. Had he instructed them to publish the papers immediately, Newton would have exposed them to potential censure and ridicule by those who were not prepared to receive the truth they contained. Had he explicitly ordered them to keep the papers secret, he would likewise have implicated them directly in his potentially dangerous beliefs. Instead he chose a middle ground whose very blurriness was a form of protection. That during his life Newton preserved his papers and his secrecy while leaving no clear directions for them after his death therefore leaves us with a deep mystery, but also a sense of Newton's awareness of both the limits of his contemporaries' understanding and the possibility of a more enlightened future. He left his papers unassigned but also undestroyed. Protected by secrecy and sheer disorder and the bonds of friendship and family, they were left to float down the byways of time.

2

# The Inheritors

Newton's heirs were a clamorous lot, and there was plenty to clamor for. The patrimonial estate at Woolsthorpe went to the closest surviving Newton, a hapless man named John, whose great-grandfather was Newton's uncle and who the local vicar described as "God knows a poor Representative of so great a man."[1] (He would gamble and drink the estate away in the course of six years.) Eight of Newton's half-nieces and -nephews, the children of his two half-sisters, Mary and Hannah, and his half-brother, Benjamin, divided the stocks and annuities, a far more substantial portion of the estate, among themselves. His nearly two thousand books, valued at £270, were sold to the warden of Fleet Prison, who bought them for his rector son for £300.

It was the "reams of loose and foul papers," stacks of rough drafts dirtied by countless revisions, that proved hardest to value.[2] What was in the papers, his heirs wanted to know. Was there money in them? At his death their kinsman was a very famous man. Surely anything associated with his name, even these old writings, could be published at some profit. A series of "contests & disputes" among the heirs concerning the disposition of the manuscripts followed.[3]

Only Catherine and John Conduitt considered the papers more than simply a source of cash. John Conduitt had briefly been a student at Trinity College, Cambridge (some ten years after Newton's departure), before becoming a judge-advocate with British forces in Portugal and eventually captaining a regiment of dragoons. John and Catherine had met by 1717 and were quickly married. In addition to his service as a member of Parliament for Whitchurch, Hampshire, Conduitt had been active in helping Newton in his position as master of the Mint

John Conduitt inherited Newton's papers on his death. He gathered reminiscences of Newton and made an unsuccessful attempt to publish a biography of his friend, patron, and relative by marriage. © Victoria and Albert Museum, London.

(a post he assumed on Newton's death). The two men had evidently grown close. Conduitt found himself, by dint of his marriage and the ties of affection, closely involved with the disposition of Newton's papers after his death.

John and Catherine were adamant that the papers should be saved both as a repository of Newton's deeper contribution to society and for the elucidation of divine order, so that "the labour and sincere search of so good a Xian [Christian] and so great a Genius, may not be lost to the world."[4] A committee of three administrators, including Catherine Conduitt, was appointed to oversee the process of deciding which of the reams of papers were worthy of publication (and thereby could raise money for the heirs). The committee brought in Thomas Pellet, a member of the Royal Society, to examine them. He spent just three days with the papers in May 1727, going through an enormous amount of material, before drawing up an account of what he found, divided somewhat arbitrarily into eighty-two separate items, which he entitled "List of the papers & manuscripts belonging to Sr Isaac Newton K[nigh]t deceased taken by the relations upon perusing and examining the same."[5] Pellet was the first person to carefully examine Newton's papers after his death. What order there was in the papers—which is not at all readily apparent from the vague terms of the inventory—was there for Pellet to see. Whether a system was discernible to Pellet is unknown. His list suggests a man trying to navigate in the dark. Descriptions range from the moderately helpful ("being loose papers relating to the Dispute with Leibniz"; "a bundle of English and Latin letters to Sir Isaac"; "being an Historical Account of two notable Corruptions of Scripture") to frustratingly vague ("being loose Mathematical papers"; "being Chemical papers"; "being loose and foul papers relating to the Chronology"). Indeed Pellet turns repeatedly to the words *loose* and *foul* to characterize the papers, an understandable response given the number of heavily worked drafts and revisions on a range of material. But Pellet was equally limited in his description of the notebooks, unhelpfully describing what is now known as the Waste Book, with its significant mathematical notes, simply as "A folio Common-place book; part in Sir Isaac's hand."

Pellet's challenge would not have been nearly so great had Newton written only in notebooks (called paperbooks at the time). But like his contemporaries he used loose sheets frequently and in many different ways. Newton usually bought what was known as "pot sized" leaves (after the watermark they carried); a piece was about 16 by 12.5 inches. Normally he used paper as he bought it, folded in half to form a simple folio. He generally wrote only on the recto, or right-hand page, inserting addenda or corrections on the blank verso sides if necessary. Sheets of paper could be attached in folded folios or with string. For many of his alchemical writings, he folded the paper in half again to create a booklet with four leaves. (He slit one side along the top edge so that the booklet could be paged from folios 1 to 3 before being opened out to reveal the fourth page.) But often it seems that whatever order was imposed was done so in the simplest way possible: by stacking or bundling. Newton was a parsimonious man. He reused fifty-year-old scraps of paper, and this meant that the jottings of early adulthood often shared the same page as those of his dotage. To make matters worse, most of the papers were undated. With the exception of a few undergraduate notebooks, proudly labeled, for example, "Isaac Newton, Trin. Coll 1661," most of the papers contained no indication of when they were written. References to contemporary events were frustratingly rare. Sheer numbers alone—which Pellet did include—give a sense of the scale of the task. There were 353 "half sheets in folio; being loose and foul papers relating to the figures and mathematicks," 495 such sheets "relating to Calculations and Mathematics," and a whopping 606 half sheets in one bundle "relating to the Chronology," evidence of the depth of Newton's interest in history.

Pellet's declarations were most assured when it came to his central task: that of determining which, if any, of Newton's remaining manuscripts were "fit to be printed" and therefore could be sold for cash on behalf of Newton's heirs. It didn't take Pellet more than three days to determine what was worth printing. Out of the mass of papers and notebooks, he considered just five documents worthy of the press.

Was Pellet conspiring to conceal secrets contained in the archive pertaining to Newton's religious and alchemical beliefs? Had he been conscripted to aid in the preservation of Newton's reputation, a matter of national pride? This has been suggested by some who came after and wondered why so little from the trove of papers was released and so much was consigned to obscurity. But it seems unlikely. In truth Pellet was merely discharging a duty to the inheritors of Newton's estate, as he had been hired to do. The question he had been asked to address was whether anything in the papers could be published as it was and therefore earn money for Newton's heirs. He declared just a single document ready for immediate printing: the *Chronology of Ancient Kingdoms* that Newton had labored to make a fine copy of in that dark room just days before his death. That was sold immediately—for an impressive £350 (£30,000 today)—and published in the following year.[6]

In the next few years only two other documents from this cache were published. One was Newton's draft of what was originally intended to be the last book of his master work the *Principia Mathematica*, published in 1728; the other was his *Observations upon the Prophecies*, a technical work in scriptural exegesis, the practice of analyzing the words of the Bible to better understand their hidden meanings, which was published in 1733.[7] Strikingly different as they are, both these writings evince a lifetime devoted to study. (The other two documents that Pellet had identified as publishable—thirty-one half-sheets in folio on "paradoxical questions concerning Athanasius" and an "imperfect mathematical tract" most likely on aspects of the calculus—remained unpublished until the twentieth century.)

Aside from these documents, John Conduitt inherited everything else, a confusing jumble of papers covered with densely written notes in Newton's neat hand, spanning well over sixty years of scholarship and analysis.

What did his heirs expect to find in these papers? The assumption that the manuscripts of a person, even as celebrated a thinker as Newton, might be worth saving is a relatively new phenomenon. It dates, not coincidentally, to Newton's own era and relates to the changes in scholarship and knowledge in the period, when exploration opened new worlds to investigation. As the search for new things widened, the need to preserve the new knowledge intensified. New systems for organizing knowledge arose, from plans for a universal language (Newton took notes on one) to shorthand writing techniques (useful for both secrecy and efficiency), from cabinets of natural history that physically arrayed new specimens, to abstract schemes, such as Francis Bacon's *Novum Organum*, published in 1620, or the German British polymath and "intelligencer" Samuel Hartlib's private, and massive, compendium of natural knowledge.

In this, as in so much else, John Aubrey, author of the gossipy and sharply observed compendium of so-called minute biographies that is best known by its Victorian title, *Brief Lives*, is an entertaining guide. Aubrey was one of the first men who thought that preserving the manuscripts of great men made sense and, indeed, was morally and philosophically requisite. Frustratingly for him, he found his fellow men insensitive to these matters. Nonetheless he worked indefatigably to conserve papers, objects, portraits, and anecdotes that would not only reveal the achievements of the important men of his day but leave a vivid impression. Aubrey was vocal in his dismay that others did not recognize the urgency and importance of his project. He worked to preserve, among others, the manuscripts of such men of science as Francis Bacon, Thomas Harriot, and the astronomer and mathematician John Dee. Aubrey's curiosity was boundless, and he simply could not understand why others did not share his profound desire to know—and through their papers and manuscripts to preserve—what made other men tick. "Oh Anth[ony]," he wrote his friend Anthony Wood, lamenting the dispersal of great libraries after their owners failed to specify what should become of them after their death, "what worke doe the Executors and widows make with Librarys which were so dearly beloved by their late Masters! And Oh that men would be but more publique spirited and make their hands their Executors and their Eies their Overseers."[8]

Aubrey frequently counseled the Royal Society to keep or publish the papers of particular scholars, and he himself donated books to the Royal Society library and specimens to its museum, as well as books to the Bodleian Library, the Ashmolean Museum, and the New Inn Hall, Oxford. But it was never enough, and he frequently bemoaned the loss of yet another valuable anecdote or bit of knowledge that men had idly allowed to go unnoted. Aubrey seems to have passed through life with an exquisite sensitivity to both the value and the evanescence of the things he encountered, the ease with which the records of even the greatest of human achievements could be lost.

He lamented that he had had to rescue the papers of the mathematician and bishop of Salisbury Seth Ward "from being used by the Cooke," who had planned to place them under cooling pies. It wasn't just cooks that were to be feared. Aubrey adumbrated the risks of "the good huswives" who might use papers for all manner of sacrilege, including to wrap herring, or else sold "by weight to the pastboard makers for wast paper," or used to load firearms or put under "the tayler's sheeres."[9] From Aubrey we know how common it was for men to be disdainful of their own papers, neglecting to inform their executors how to preserve them. Newton not only saved an enormous amount of paper, but that material has survived, more or less intact, to the present day. How it happened that Newton's papers survived is, like most things historical, a matter of both luck and determination.

The terms under which Conduitt inherited the papers were complex. Newton had died while holding the post of master of the Mint, which in those days required that its holder assume personal responsibility for the probity of each new coinage of money. That meant that at Newton's death he had nominal debts amounting to the entire sum of Great Britain's national coinage. John Conduitt agreed to take on this debt until the coinage had been certified, accepting liability for any imperfections in the coins. In exchange for assuming this risk, he asked for, and was granted, Newton's manuscripts. This must have seemed a good deal to the other heirs, to whom the papers had already been certified by Pellet as more or less worthless. Conduitt also agreed to post a £2,000 bond as a guarantee that the heirs would benefit should any material be published in the future.

Like Aubrey, John Conduitt's own mind ran to biography, and he venerated the man who had become his uncle by marriage and who had served as a friend and supporter. In the months following Newton's death, Conduitt mobilized Bernard de Fontenelle, the renowned secretary of France's equivalent to the Royal Society, the Académie des sciences, to write a fitting eulogy to Newton. During Newton's lifetime, the French had been resistant to his innovations. What better way to resuscitate Newton's reputation on the Continent, while consolidating it in England, than by inviting a friendly Frenchman to praise Newton?

Just one week after Newton's death, Conduitt wrote several letters to Fontenelle, asking him to formally undertake the task of eulogizing his English counterpart and adding that he was preparing his own version of Newton's life. Fontenelle wrote back asking for biographical details. Were his talents apparent in boyhood? What were his favorite books? How did he die? Conduitt larded his response with unabashed superlatives. Newton was, Conduitt wrote, "pure and unspotted in word & deed," "exceedingly courteous and affable even to the lowest and never despised any man for want of capacity," and with "such a meekness and sweetness of temper that a melancholy story would often draw tears from him." He was deeply opposed to cruelty to man or beast, "mercy to both being the topick he loved to dwell upon." And finally, Newton was exceptionally charitable to others and "always lived in a very handsome generous manner thou without ostentation or vanity, always hospitable and upon proper occasions gave splendid entertainments."[10]

Though in addition Conduitt supplied Fontenelle with some copy to help him on his way, he was not satisfied with the published elegy. Ultimately, in Conduitt's view, Fontenelle had damned the great Newton with faint praise, due, perhaps, to a regrettable loyalty to France. According to Conduitt, Fontenelle had demonstrated "neither the abilities nor inclination to do justice to that great man who had eclipsed the glory of their Hero Descartes."[11]

Conduitt sought to do better: to provide a lasting monument to Newton in the form of a biography. And indeed he was the first to look at Newton's papers with an eye toward what they could reveal of the man. (Pellet had been scanning, and quickly, for print-ready copy.) He did not exhaustively catalogue the papers, though he did describe the contents of some of them and listed some of the different paperbooks that Newton had preserved.[12] Instead of comprehensiveness he sought the telling anecdote, the biographer's gold, to brighten the mass of detail with something fresh and surprising. His eye was good, and much of what he found remains of interest today. He also spoke to many of Newton's contemporaries and, before his death, to Newton himself, with the aim of building a proper biographical monument to the man.

Biographies were a rarity and a novelty at the time. The only lives deemed unquestionably worth studying were those of the saints. Conduitt felt the need to justify such a project, and his apologia makes it clear how unusual his idea was. Since even historians of the Roman Empire wondered about the worthiness of their accounts of the great rulers, Conduitt admitted, some might think it "a dry and unaffecting employment to compile the life of a private man, spent in speculation &

in the exercise of those silent virtues which, however delightfull to the possessor, afford but little entertainment in the description, & are not so apt to strike a vulgar reader as the tumultuary scenes of pomp & action."[13] There was no way around it: Newton had not led a life of action or adventure. His was a reserved life, full of quiet study and hopelessly bereft of public spectacle. It did not make for obviously gripping reading.

Conduitt offered a novel reason for this new kind of biography, one that related to the broader growth in knowledge (now known as the Scientific Revolution) to which Newton contributed so much. He suggested that understanding the "particle of Divinity" achieved by a man such as Newton was as pleasing to the mind as "following a Conqueror through a field of blood and confusion." Newton was a better role model than the leaders of violent struggles. His life of "labour, patience humility temperance & piety without any tincture of vice" provided a much better, more "universally beneficial" example than tales of a Caesar or an Alexander. But piety alone didn't qualify Newton for a biography. As a natural philosopher, he was the prime progenitor of a radical new system of thought, and that made him something of a "speculative Conqueror" who had extended the limits of human understanding.[14] Newton had discovered new things of such astonishing scope that he had earned a place in the pantheon of great men.

Conduitt preserved in his notes the reminiscences of those who knew Newton best. He gathered his anecdotes from a wide variety of sources, including Newton's nephews, luminaries such as the Earl of Halifax, Dr. Arbuthnot, and Richard Bentley—who had exchanged letters with Newton, pushing him to explain the religious implications of his theory of gravity—and Catherine, his wife and Newton's half-niece.[15] Conduitt had interviewed Newton himself in his later years, recording memories that Newton recalled from his childhood some seventy-five years earlier. These stories form the core of every tale that has been told about Newton's life and personality since. Conduitt included what had to be the oldest story about Newton: that of his birth and unlikely survival. He was so little upon his birth that he could have fit inside a quart pot and so weak that the women sent to announce his birth "sate down on a stile by the way & said there was no occasion for making haste for they were sure the child would be dead before they could get back."[16]

From such unlikely beginnings Newton gradually (and, in Conduitt's telling, inexorably) rose to the height of intellectual dominance, which, once achieved, he never surrendered. Writing was, from the earliest age, a source of his mastery, according to Conduitt. Originally a poor student, Newton had risen to the top of his class in school only after receiving from another boy a "kick on his belly which put him to great pain." He bested the boy physically and proceeded to take further, intellectual vengeance. This included making himself "Master of his Pen." Newton's earliest examples of writing and sketching show his restless mind at work, and in an "old pocket book in which he has writ his name & the date of the year 1659 there are several rules for drawing and making colours."[17] Once started, Newton could not stop. Conduitt noted that the walls of his childhood home at Grantham were covered with drawings of "birds beasts men & ships well designed, & several persons remember many of his drawings both from pictures & the life," which included the head of King Charles I—who was beheaded—as well as John Donne and "his worthy schoolmaster Mr Stokes."[18]

Newton quickly developed the skills needed to excel not just at school but at university. Perhaps the most important was his lifelong habit of incessant note-taking. Newton was "hardly ever alone with out a pen in his hand & a book before him."[19] This practice began with his first undergraduate notebooks and continued throughout his Cambridge years, notably in a small notebook in which he recorded a list of "certain philosophical questions," a framework for research that lasted him throughout his life. He continued to refer to this juvenile notebook even decades later. More generally the skills he acquired in this early exercise in research provided the foundation for all of his investigations—in natural philosophy, in alchemy, in theology, and in Church history. His abilities weren't limited to his tremendous mathematical skill or his far-reaching insights into physics and optics but encompassed something more fundamental: the capacity to critically interrogate a text and take notes that formed the basis for his own, innovative research. Newton's note-taking practices, acquired in the 1660s, were still novel enough in 1727 that Conduitt thought it worth describing how Newton "used to write down any thought which occurred upon the books he was reading, & make large abstracts of them." From the beginning Newton was patient, persevering, and productive. Evidence for his lifelong studiousness was contained in the "the rheams of foul and loose papers he has left behind in his own hand...some of which are the same thing writt over six or seven times."[20]

The multiple copies of the same text that Conduitt found amid the reams of papers demanded explanation. Fully committed to portraying

Newton as exemplary in all ways, Conduitt interpreted these puzzling documents as evidence of Newton's virtuous perseverance in pursuing—and perfecting—a line of inquiry over many iterations. It was a noteworthy aspect of the papers to which future commentators would return, sometimes with less positive reactions.

Alongside stories about the proliferation of manuscripts, Conduitt supplied tales of their destruction. He related how once Newton had left a candle on some papers on which he was working and went out to meet someone, only to return to find that the candle had set fire to the papers, which included his work on mathematics and optics.[21] As warden of the Mint (a position Newton held for four years before becoming master, which was a loftier position), Newton had attended all the trials of the "clippers and coiners," men who had counterfeited money. In the process he had generated a great deal of paperwork and, Conduitt reported, had helped to burn "boxfulls of informations in his own handwriting" relating to that work.[22] Finally, Conduitt's portrait of Newton recounted how the great man had retained his full faculties right up until the moment of death, including his ability to write with a "small but very distinct & legible hand" that remained steady to the last.[23]

Conduitt also made a stab at cataloguing and editing some of the papers. He took extensive notes on Newton's draft of "The Original of Monarchies," a treatise consisting of seventy folios, listing the contents of each page and noting (as modern editors do) where Newton had deleted or added material. Conduitt's sensitivity to the importance of these additions and deletions as evidence of the progression of Newton's thinking would be unmatched for centuries. He also demonstrated a thoughtfulness about publication. Though Newton had written much that was of interest simply because he had authored it, not everything deserved the press. To help him (and his unidentified judges), he drew up a chart. On one side were reasons to publish; on the other side, reasons not to publish. Militating against publication were the facts that some of the material had already appeared (in the *Chronology*) while some of the unpublished material remained, as Conduitt put it, "very Imperfect." In favor of publication was the fact that much material, in good form, had not yet been published. More to the point, this was Newton's writing, and therefore, "however imperfect some of these papers may be, yet certainly they must contain something very valuable to the publick."[24]

Conduitt's reasoning—that anything Newton wrote must be valuable—was compelling. "The Original of Monarchies" was indeed published soon after he wrote those words. But with a few exceptions, nothing else from the jumble of papers would be made public for nearly one hundred years, and even today, nearly three centuries later, most of it remains as obscure and unknown as it was in Conduitt's day. Conduitt was as close to Newton as anyone, and as committed to the idea of a biography as a lasting testament to an English hero, and yet he failed to complete—or even make much headway on—his projected biography. The notes he left are without equal in their evocation of the man, but they are as unfinished and sketchy as notes can be.

Conduitt didn't need to fully organize the papers to understand what they contained: nothing less than a lifetime's worth of deeply impassioned investigations of Christianity. It was complex and contentious material, but it boiled down to this: Newton believed that in the fourth century A.D. early Church fathers had inserted the pernicious fiction into Church doctrine that Christ was an equal partner in the Holy Trinity. Newton argued that a full and true history of the Church would reveal what most had forgotten or never knew: that Christ was subservient to God the father. This belief was called anti-Trinitarianism because it denied the Holy Trinity of father, son, and holy ghost—a denial that was heretical to mainstream Anglicans of the day.

What would it mean if the father of Enlightenment reason were to be revealed as an obsessive heretic whose denunciations of modern Protestantism were full of violent, bloody imagery, suffused with the hatred and viciousness of a man who believes that the salvation of humanity is at stake? Newton had been a Christian, and he had certainly been a "Genius," but neither Conduitt nor any of the others involved in sorting his papers had any appetite for revealing the full extent of his highly idiosyncratic form of Christianity. These were the kinds of beliefs that landed heretics in prison or, in exceptional cases, saw them executed. Though the last people to be sentenced to death for anti-Trinitarian heresy in England were burned at the stake in 1612, in Scotland Thomas Aikenhead was hanged in 1697 for expressing heretical views on the incarnation of Christ.

But while the archive certainly contained items that could embarrass and perhaps even horrify Newton's supporters, both the range and the disorder of the papers protected them. There were few enough men who could read and understand the *Principia*. The number who could manage the geometrical language of that work, the historico-chronological references of some of his religious writings, as well as the mixture of quantitative and allusive terms that appeared in Newton's chemical writing was smaller still. On top of it all, the archive itself spoke in its own language of confusion, of shuffled, misplaced, and canceled pages. If it was decipherable at all, that kind of language required a mind of still finer discrimination to unravel it.

There was likely not one but many things that defeated Conduitt in his bid to write a biography worthy of Newton: the complexity of the material, the pressure to produce something worthy of the man, the bond securing profits for the heirs, and the presence in the papers of material that could embarrass Newton posthumously. It all added up to a mess that was easier kept out of sight.

As the inheritor of the papers and the man who had nominated himself to be the prime mover in consolidating Newton's posthumous reputation, Conduitt had the motivation and the means to keep information private. He kept the papers in his house and away from prying eyes. When he died in 1737, Conduitt was buried in Westminster Abbey, on Newton's right-hand side. Catherine joined him there after her death two years later. In death, as in life, the Conduitts remained close to Newton, sharing, by proximity, in the honor of the memorial to his great accomplishments. Elsewhere, Newton's papers and the evidence they contained of a man of complex belief and profuse interests, were secured out of public view, awaiting a time when their contents might be deemed more palatable.

3

# Petrifying Newton

It is tempting to treat the decades that followed Newton's death in 1727 as a time when his supporters fortified his reputation as a man of saint-like piety and nearly divine intellect. This is the story Conduitt sought to tell, a story of marmoreal rest, with Newton at ease in his monument in Westminster Abbey, furls of stony fabric unfolding around him like an eternal dream of adoration, his reputation stamped permanently, like some honorific medal. By this telling the manuscripts Newton left behind sat like a suppressed remnant of his secret self, the dark side of a great man, hidden away by his sworn defenders.

But the real story is different. Eulogies were composed, medals struck, and poems written to mark the passing of a great thinker and a great Englishman, but the petrification of Newton's reputation didn't occur right away. In the first years and even decades after his death it was fairly widely rumored that the great man's beliefs had shaded into the further reaches of radical Protestantism, if not outright heresy. Whatever his papers contained, it was assumed, only augmented what was circulating in quasi-public spheres about him. They were like an additional daub of paint, adding definition to a picture whose outlines were there to see for those who cared to look.

Newton's heresy was known during his life to those who could read the signs, some left by Newton himself. He himself had added an appendix to the second edition of his greatest and best-known work, the *Principia*, published in 1713, called the General Scholium, in which he described the kind of God his universe required: everywhere and omnipotent. Such a divinity left little room for anything else, including a Trinity. If God was already everywhere, how could Christ also be fully divine? For those who were paying attention, Newton had left a significant clue to his anti-Trinitarianism hidden in plain sight.

In his social life too Newton had left markers to his radical beliefs, in the form of his friendships, or association, with men such as the philosopher and clergyman Samuel Clarke and especially the mathematician and theologian William Whiston, both of whom publicly aired strongly heterodox beliefs. In 1712 Clarke, one of Newton's most trusted confidants, published a book in which he argued that there was no support for the doctrine of the Trinity in the Bible. Whiston went further, sacrificing his career as a Cambridge professor to publicly preach an even more explicitly anti-Trinitarian doctrine. That Newton was known to consort with such men was itself a powerful indictment of his orthodoxy. Whiston had been bold enough to publicly hint at Newton's beliefs during Newton's own lifetime.[1] More moderately, as intimate an acquaintance as Newton's physician, Richard Mead, thought that he was a Christian and "believed revelation" but admitted that Newton had perhaps not believed in "all the doctrines which our orthodox divines have made articles of faith."[2]

Much, then, was known, or guessed, about Newton's beliefs during his lifetime. Much was also ignored, out of a combination of denial, lack of interest, and the desire to protect the great man's saintly image. After his death Newton's anti-Trinitarianism for some years became an even more openly debated topic among those whose impulses to reveal had been held in check by the weight of his public persona and who felt themselves freed by his death to use him for their own devices.

Several contemporaries wrote to John Conduitt immediately following Newton's death, expressing the hope that his religious writings could now be published in order to vindicate Newton, not as a scientist but—in the face of an increasingly irreligious Enlightenment—as a man of God as well as intellect. The mathematician and Newton's longtime friend John Craig hoped the papers would remind people that Newton had studied nature in God's name: "They were little acquainted with him who imagine that he was so intent upon his studys of Geometry and Philosophy as to neglect that of Religion." Although he acknowledged that Newton's religious thoughts were unconventional, Craig nevertheless urged that they be published to redeem Newton in the eyes of "infidells" who claimed that Newton's study of religion was limited to his infirm dotage. "But now its hoped," wrote Craig, echoing the Conduitts, "that the worthy and ingenious Mr Conduit will take care that they be published that the world may see that Sir I: Newton was as good a Christian as he was a Mathematician and Philosopher."[3] For Craig, Newton's beliefs were not heresy but proof of his piety.

Others were more concerned that Newton's reputation remain unsullied by anti-Trinitarian associations, even if this meant forgoing proof of his piety. The Scottish clergyman Robert Wodrow also inquired about the status of papers that "Sir Isaack left behind him." Did any bear on religious matters? Wodrow was relieved to hear from Colin Maclaurin, a Scottish mathematician and friend of Newton, that "there was nothing they had seen as to that, or any other subject in Divinity," since Wodrow seemed to know that Newton's beliefs were unorthodox. And Newton's influence, even posthumous, was significant enough that "any small innuendos from a man of his character upon subjects, it may be, he had not thoroughly considered, would be very much improven and sualloued doun by multitudes."[4]

Of course, Newton's papers contained much more than "small innuendoes," but that didn't matter so long as the papers were kept away from prying eyes. Conduitt himself cannily acknowledged that although Newton was a firm believer in revealed religion, as the papers showed, his "notion of the Christian religion was not founded on a narrow bottom."[5] This could be read in two ways: that Newton was generous in his interpretation of the range of others' beliefs or that he was generous in including his own aberrant interpretation within the fold. Ever the diplomat, Conduitt found a way to make a virtue of the dangerous distance that loomed between orthodoxy and Newton's beliefs.

Not everyone was convinced. Two years later Maclaurin revealed (again via Robert Wodrow) that a number of papers about "Scripture Prophesys" that Newton had left behind did indeed contain some "peculiar thoughts" on prophecy and, tellingly, that he himself had heard Newton "express himself pretty strongly upon the subordination of the Son to the Father." For those committed strongly either to denying or maintaining the anti-Trinitarian stance, the publication of the great scientist's views was eagerly anticipated. "There is a great expectation in England," reported Wodrow, pertaining to revelations about Newton's papers on the prophecies.[6]

These expectations were partially met by the publication of two documents. Again, depending on where people fell on the matter of anti-Trinitarianism, the printing of the *Chronology* in 1728 and the *Observations* five years later brought either opportune or unwelcome new attention to the nature of Newton's beliefs. Neither work contained outright heresies, but to the sharp-eyed reader they contained unmistakable hints about how his view of the Christian past differed from the mainstream. If there was little obviously heretical about the *Chronology*, even the dedication of the *Observations* was telling. The book was dedicated to Peter King, a known anti-Trinitarian who had served as an intermediary between Newton and King's cousin, John Locke. Newton's association with Locke was important because they had exchanged radical interpretations of scripture, interpretations that, had they been made public, would have eliminated any doubt about the extent of either man's heterodoxy.[7] What Newton shared with Locke was a detailed "Historical Account of Two Notable Corruptions in Scripture," in which he used references to ancient manuscripts to document how key passages in John and Timothy, normally taken to support the doctrine of the Trinity, had been corrupted by errors of transcription and translation over the course of hundreds of years. Though Newton had initially asked Locke to have the treatise published anonymously, he soon changed his mind, and the document was not published during his lifetime. The anonymous manuscript continued to circulate, however, and some, including Whiston, had guessed at its true author.

In the *Observations* Newton undertook a forensic analysis of the Bible, the aim of which was to understand the true referents of the prophetic images and thus to locate evidence of divine providence in the history of the world. There was no such thing as "mere" historical interest for Newton. The earthly history of human events from the birth of Christ onward was indissolubly linked to the biblical prophesies that gave evidence of divine will in the past as well as of the inevitable unfolding of that will in the future. The practice of decoding which of the divine prophecies included in the Bible had already been fulfilled—by referring to historical events that matched the descriptions in scripture—was a well-accepted one. But the books of Daniel and Revelation in particular were notoriously difficult to interpret, dense with symbols and bizarre images whose historical counterparts are anything but clear. Newton made these abstruse books his specialty, seeking to unravel the history of the Christian world from the phantasmagorical assemblage of beasts and dragons that they contained.

To Newton, the difficulty of the images was evidence of their importance. He believed that ancient peoples had spoken not in ordinary language but in a thick paste of symbols and poetic speech just like that found in Daniel and Revelation. The wild and dramatic happenings they describe—the opening of the seven seals, earthquakes, famines, and meteor strikes—stood for events that were either destined to occur in the future or had already occurred in the past. The task Newton set himself in the *Observations* was to match the descriptions of events, such as the opening of the seven seals and the arrival of the Whore of Babylon, to actual events in history. By carefully combing through the scriptural descriptions, he could then determine how many prophesies remained to be fulfilled before Judgment Day. Boldly reordering the highly imagistic descriptions and using standard interpretive assumptions (taking a day in prophecy to mean a calendar year, for example), Newton was able to work out a timeline. Though the chronology of prophetical events could be used for prediction, he shied away from the vulgarity of assigning dates to future events. God had communicated the prophecies "not to gratify men's curiosities, by enabling them to foreknow things; but that after they were fulfilled, they might be interpreted by the event; and his own Providence, not the interpreters', be then manifested thereby to the world."[8] For Newton, the history provided evidence of God's providence in the past, and that was more than enough.

Reaction to the publications was mixed. Then as now, people saw what they wanted to see. Voltaire, whose religious beliefs (or lack thereof) were also suspect, explained Newton's work on the *Chronology* as a kind of mental vacation from the "fatigue of his severer Studies," which nevertheless still displayed "some marks of that creative Genius with which Sir Isaac Newton was inform'd in all his Researches."[9] Sympathetic both to Newton's ambition to reform chronology and to his controversial conclusions, Voltaire nevertheless noted that Newton "appears to us stronger when he fights upon his own ground," using astronomical observations rather than historical assumptions as the basis for his revisions.[10]

But Voltaire was exceptional in holding himself above the fray. Most who responded did so as proponents (often passionate ones) of one side or the other in the matter of anti-Trinitarianism. Newton does not seem to have changed many minds (as the defenders of contemporary orthodox Christianity feared his "poisonous" views might). Instead his views served, not for the first time and not for the last, as ammunition in an ongoing theological skirmish.

Arthur Bedford, a chaplain who had written a treatise declaiming the "horrid blasphemies and impieties" to be found on the English stage, responded with defensive outrage. Beleaguered by what he saw as a dangerous moral drift, Bedford considered Newton's failure to agree with the dates given by orthodox writers as evidence of a dangerous general tendency to undermine Christian authority. "We live in an Age, when we cannot be too cautious," he warned. "The *Divinity* of our blessed SAVIOUR is struck at by the Revivers of ancient and modern Heresies."[11] Newton was among the guiltiest of these Revivers, the creator of a chronology more than a thousand years out of sync with orthodox readings of scripture, a work that was so toxic that it "ought not to go abroad into the World without an Antidote; or rather it should not go abroad at all, lest the Antidote should not be strong enough for the poison."[12]

Too late (the *Chronology* was already published), Bedford pleaded for Newton's wishes regarding publication to be respected. Newton himself informed us, he wrote, that his treatise was "written without any Design to publish it." According to Bedford, Newton knew perfectly well that it was impossible to prove the truth of his chronology in the way he had proven his natural philosophy. More significantly, argued Bedford (with a note of rising desperation), Newton himself knew that his *Chronology* "would engage the World in fresh Controversies."[13]

Others responding to Newton managed to strike a tone that was simultaneously dismissive and vicious, referring to the distasteful heresy called Arianism, named after the early Christian presbyter Arius (250–336), who argued that the Son of God was created by, and therefore inferior to, God the Father. Arius had been pronounced a heretic in the fourth century, and from then on any association with him was anathema to mainstream Christians. For some, Newton's beliefs strayed uncomfortably close to those of Arius and his followers. Daniel Waterland, the master of Magdalene College, Cambridge, complained of Newton's *Observations* that the "*Prophetical* Way of managing this Debate on the Side of *Arianism* is a very silly one, & might be easily retorted. But besides that, what *Sir Isaac* has said, is most of it *false History*." Waterland "scribbled the *Margin* all the way" as he read the work, making note of Newton's failings as a historian.[14] Newton was justly renowned as a natural philosopher but should have stuck to what he knew best instead of going into other matters, "where He was plainly out of his element, and knew little of what He was talking about."[15] Here was the counterargument to John Conduitt's claim that everything that Newton wrote, including the nonscientific material, was worth knowing about because Newton wrote it.[16]

One likely outcome of these publications—and the fresh controversies they generated—was Catherine Conduitt's decision in 1737 to add a codicil to her will indicating that she hoped to publish the more carefully prepared theological writings in her and her husband's possession "if God granted her life." Considering the possibility (as would so many potential editors of the Newton papers) that she might be "snatched away" before she had the chance to undertake the work, she instructed the executor of her will to send them to Dr. Arthur Ashley Sykes, a prolific writer on controversial religious topics who had publicly supported Samuel Clarke, so that he could publish them and keep them from being "lost to the world."[17] Sykes was a telling choice, since by 1737 he had already demonstrated himself to be a provocateur of no small energies, having openly supported a slew of well-known anti-Trinitarians. What would he make of the theological writings he inherited? Motivated by his devotion to the anti-Trinitarian cause, would he find more material fit to publish?

The answer eludes us, because while John and Catherine Conduitt both died within a few years of this codicil, Catherine's wishes were not immediately carried out. In 1740 her daughter, Kitty Conduitt, married John Wallop, Viscount Lymington, a member of the Portsmouth family, thus merging Newton's line of descent with an old and distinguished family. Their son, also John Wallop, would become the second Earl of Portsmouth and inherit the Newton papers. He passed along the theological papers assigned to Sykes only in 1755, probably prompted by the publication of an unauthorized 1754 edition of Newton's letters to Locke (misleadingly titled *Two Letters of Sir Isaac Newton to Mr Le Clerc*). The divine controversialist did not live to see anything from the papers into print. Sykes died later that year, and the papers were inherited by a decidedly less outspoken figure, Reverend Jeffrey Ekins, dean of Carlisle Cathedral. Unsurprisingly Ekins and his heirs kept the papers largely out of view until 1872, when they donated them to New College, Oxford, where they remain today.[18]

Meanwhile the bulk of the material, including most of Newton's theological writings, stayed with the Portsmouth family, the Conduitts' descendants. They lived at Hurstbourne Park, the Earl of Portsmouth's seat in Hampshire. The family was jealous of their privacy and seem to have been somewhat burdened by their Newtonian inheritance. John Conduitt had given a bond of £2,000 to the seven nearest kin of Isaac Newton against the risk that unauthorized copies of the manuscripts in his keeping be made, robbing the heirs of potential income. The risk of forfeiting that substantial deposit would have weighed heavily on any decision to grant access to the papers. There were many reasons for the Conduitt family to resist sharing their famous relative's papers. Whether because they feared losing the bond, wished to conceal

View of Hurstbourne Park, the seat of the Portsmouth family and location for many years of the Newton papers, in 1783. Courtesy of Hampshire Library and Information Service.

their controversial contents, disdained contact with unfamiliar scholars, or simply hoped to avoid the hassle of sorting the mass of papers, they did not make it easy for scholars to access them. Some thirty years after Newton's death, Lord Portsmouth tried to put off an inquirer by complaining that the Newton materials in his possession "were very voluminous & it will be a matter of much time & trouble to examine them."[19]

A very few managed to penetrate the familial carapace. In 1775 Samuel Horsley, a scientist, parliamentarian, and scholar as well as the secretary of the Royal Society, submitted a proposal to "publish, by subscription, a compleat edition of all the works of Sir Isaac Newton, with notes and comments, in five vols, in quarto." Horsley, a vicar who staunchly supported mainstream Anglicanism, almost certainly did not yet know what Newton's private papers contained. The Society, at any rate, backed Horsley's plan unanimously and gave him their full authority to proceed with a project that was, as the notes indicate, "highly honourable to the nation, and of the greatest importance to Science." They authorized him to have access to all the relevant papers held by the Society and directed the librarian to assist him.

Samuel Horsley visited the Newton papers at Hurstbourne Park in 1777 and published almost nothing of them in his *Isaaci Newtoni Opera quae extant Omnia* of 1779–85. © National Portrait Gallery, London.

By 1777 Horsley had managed "with great difficulty" to get access to the papers at Hurstbourne Park. This was quite a coup, and one he boasted of to his printer, William Bowyer. When Horsley got to the estate he found the papers sitting in the drawers of a cabinet and three bureaus, "somewhat randomly tied up in bundles." He prepared a rudimentary catalogue of the papers, returning a few weeks later to look more closely at the mathematical papers. He also inserted slips of paper in among the manuscripts, with notes in his own hand regarding their fitness for publication. Like Thomas Pellet, who had examined them thirty years before, Horsley resorted to calling certain of the papers simply "loose" or "foul," though he made some attempt to identify among the calculus manuscripts "a Paper on the quadrature of trinomial Curves with some Fragments relating to Fluxions."[20] But Horsley wasn't being simply vague. Some of his silence on the contents of the papers at Hurstbourne Park was due, no doubt, to his ignorance of the subjects they contained, some to what would have been their obviously anti-Trinitarian contents, and yet more to the simple fact that Horsley was under pressure from subscribers to get his publication out quickly. In his report to the Royal Society, Horsley suggested just nine items from the Portsmouth Papers to be published, all of them relating to Newton's geometrical, mathematical, and optical research. The Royal Society Council agreed and noted in return that the publication of these papers may be of "great use to Science," thanking the Earl of Portsmouth for sharing them.[21]

Horsley's eventual publication of five volumes between 1779 and 1785 did not bring much that was new to light and only sparingly drew on the texts he'd seen at Hurstbourne Park. (He used the original manuscripts to make improvements to certain works that had previously been based on faulty copies.) This seems somewhat strange, considering Horsley's unique access to the Portsmouth papers. Some of this might be blamed on a tight publication schedule and the lateness of his trip to Hurstbourne, but the truth is that the wind seems to have gone out of Horsley's sails, most likely because of the papers' heretical contents. The private papers that Newton had left behind contained within them the echo of the living man. But reanimating the man from bits of paper would have been a huge and risky task, and the conservative Horsley, who in 1788 became a bishop in the Anglican Church, would have viewed Newton's anti-Trinitarianism with extreme distaste. He was not the man for the job. As it happened, Horsley's association with the Royal Society came to an end in late 1778, even before his Newton edition had appeared in print, when he abruptly surrendered his secretaryship and was dropped from the Council. Horsley's edition would stand as the high point of his Royal Society tenure.

And so, despite Catherine Conduitt's desire that additional material from Newton's theological writings be published and despite the interest in Newton's religious views among a group of active Anglican divines and their radical counterparts, there was precious little new Newton material that surfaced in the eighteenth century. As the century wore on, the living memory of Newton as a complex human being, a man whose beliefs were not, as John Conduitt had acknowledged, "founded on a narrow bottom," was lost. In its place emerged a smoother, more mythical Newton, whose rougher edges had been worn down by time. This was a man of semidivine insight whose countenance was unshadowed by any hint of heresy.

Newton entered the nineteenth century like a petrified mummy, all wrapped up and immune from the putrefying forces of history. His name and works had been sanctified by monuments, his earthly frailties lost from memory. Despite the scraps of evidence about his religious beliefs that blew hither and yon in the eighteenth century, visible to those who knew where to look and cared to do so, little information that was not favorable to Newton made it into broad circulation. Newton had been saved for eternity, his promoters hoped, in the gloomy quiet of the temple to science that he himself had helped build.

4

# The Madness of Newton

It wasn't just Newton's stash of private papers, watched over by the Portsmouth family, that had been hidden away during the eighteenth century. Traces of Newton existed in all sorts of other places: in the collections of letters of men with whom Newton had corresponded and in the archives of institutions, such as the Royal Society and the Royal Observatory at Greenwich, that contained records of what Newton had said and what he had done. This extensive documentation—which suggested a richer story of Newton's life than the one John Conduitt authorized—had remained unknown to all but the most dedicated and sophisticated of analysts throughout the course of the eighteenth century. But in the nineteenth century came a passion for uncovering the past. Digging up the dead and their paraphernalia was all the craze. What had lain quiet for a century was beginning to be disturbed by the scientific spirit of the age, which sought objectivity and truth in history no less than in astronomy or physics. Napoleon's invasion of Egypt brought news of soaring pyramids and bejeweled tombs, relics of an ancient civilization of unimaginable scope. Mary Anning found dinosaur fossils along the coast of Dorset, evidence of a history of the earth longer and stranger than previously believed. And the hungry digging of railway cuts laid bare the strata of previous ages like the pages of a book. The fashion for biographical exhumation grew too as new scraps of archival information surfaced from cupboards and attics throughout the country.

By the 1820s Britain had started to emerge from the preoccupations of the Napoleonic wars and the Industrial Revolution began to rev into high gear. The nature of Newton's life, and of his putative genius, became, like the butterflies that growing numbers of amateur naturalists marveled at under their microscopes, a subject worthy of attention for the rising middle classes. With the effects of technology visible as never before—in bridges and railroads and factories—the role of science in society took on a new urgency. Interest in Newton as the founding father of modern British science, the man who had unlocked the laws of the universe and opened the door to their exploitation, grew correspondingly.

Since his death Newton had served both children and adults as an example of the nearly unlimited potential of human achievement. The archival material gradually but inexorably reemerging about him in the beginning of the nineteenth century challenged what had become the comforting truisms of three generations and generated new reasons for studying his life and personality. Rather than resting secure in his status as Britain's scientific saint, Newton became a subject for vigorous debates over the relationship between intellectual achievement and morality. "Was he a *good man*," asked a child in one popular text, "as well as a great philosopher?"[1] Given the preoccupation of the Romantic generation with "genius" and its psychological aspects, it was fitting that interest in Newton would focus initially not on his religious beliefs (though anti-Trinitarianism retained the power to shock many Christians) but on his state of mind. These disputes did not take place in drab historical societies or in the salons of the elite but in the pages of mass-circulated newspapers and books, written by the leading scientists and journalists of the day and avidly read by a newly literate and emboldened public.

Previously unexamined manuscript sources played a crucial role in this debate. Newton's own private papers were capable of producing startling revelations long after his death, helping to prolong precisely the sort of litigiousness in matters of scientific reputation that Newton had lamented in his own time (while stubbornly fighting his corner to the bitter end). New information gleaned from additional sources, such as the letters and diaries of Newton's contemporaries, including John Locke, the Astronomer Royal John Flamsteed, and Samuel Pepys, extended the scope for debate over the sort of man Newton had been. If Newton's lawyerly concern for evidence seemed vindictive (or merely tenacious) in the controversy over the calculus and the theory of colors, for example—two fiercely contested subjects during his own lifetime—by the nineteenth century a passion for uncovering what had really happened became widespread. Even Newton's most dedicated followers welcomed the chance to present their case for his genius. The metaphor was legal: the evidence, in the form of previously unseen documents, must be arrayed for the interested reader to make his or her own judgment.

But the new documentation generated a very uncomfortable paradox. Geniuses in science uncovered universal rules of nature, but they themselves often seemed to exist beyond the normal rules of human understanding and social interaction. Was Newton an exemplary human being, or was he a superhuman whose inspiration and vision were inaccessible to even the most diligent and committed of scientific workers? The answer mattered not simply for moral but for social reasons. If Newton's mind had been ungoverned by rules, then it might actually be dangerous to continue recommending him as a paragon of rationality. The role of science in maintaining order at home and establishing supremacy abroad was at stake. In the decades following the French Revolution, it remained to be seen whether Britain too might tip over into anarchy and mob rule. Newton's achievements were a stirring symbol of human potential, but if his genius lacked method, and indeed had led to madness, then he was far from a welcome example.

The first to question Newton's sanity was a Frenchman named Jean-Baptiste Biot (1774–1862), who published a provocative portrait of Newton in a French encyclopedia. Though it was brief, merely sixty-six pages, the account contained the inflammatory suggestion that the great Enlightenment sage and paragon of English rationality had indeed gone completely and, in certain senses, irrevocably mad in his fiftieth year of life.[2] It was bad enough to state that Newton had had a period of insanity, but Biot went further. He implied that Newton's faith was a companion only of his old age (his senility, was the whispered implication). God and science had not coinhabited the great man's person when he was in prime, but only after he had completed his great designs.

That Biot should make what many considered an unforgiveable attack on Newton's moral probity is surprising, for in many ways Biot was a consummate Newtonian of the kind that France excelled in producing at this time. He was part of a group of French mathematicians led by Pierre-Simon Laplace who deliberately sought to carry the great English natural philosopher's program forward. These French *savants* applied Newtonian theories of force and motion at both ends of the scale, to large objects in the heavens (the study of which was termed *celestial mechanics*) and to very short-range forces such as those governing

In 1829 Jean-Baptiste Biot published evidence that Newton had temporarily lost his mind. Courtesy of the Smithsonian Libraries, Washington, D.C.

capillary action. Biot, as a student of Laplace in the late 1790s, had proofread his teacher's great work, *Mécanique céleste*, and duly absorbed its contents.

Biot became a leader in the drive to modernize Newtonian theory for a new age—to solve the so-called inequalities that remained in contemporary understanding of the solar system. He also contributed to the ongoing debate about the nature of light—was it a wave or a particle?—with a series of investigations into its polarization, as well as completing much original research on sound and refraction. In all this Biot was vocal in his regard for the English natural philosopher, expressing his "unbounded admiration for Newton" on an 1817 visit to Cambridge.[3]

It was therefore appreciation rather than slanderous intent that prompted Biot to write his encyclopedia article on Newton, though Biot's veneration for the great scientific work that Newton had done did not preclude his cold-eyed appraisal of the man. The encyclopedia article (then an intellectually bracing format) was published in France in 1822 to little fanfare; the French were unconcerned by the story about Newton's madness. It was not until seven years later, when the work appeared in English, that it provoked outrage in some quarters. The translation appeared in a pamphlet published by the influential Society for the Diffusion of Useful Knowledge. Publication in this format meant that Biot's piece reached the middle-class promoters of science and science education as well as the working men to whom they hoped to minister with comforting material on the benefits, material and spiritual, of education and self-help. This vision of gentle reform of the working classes was not necessarily compatible with Biot's unflinching version of Newton's life. To English reformers promoting Newton as an exemplar of hard work and diligent study, even a temporary derangement of reason was unacceptable.

The question was how accurate was Biot's accusation. He hadn't managed to gain access to the Newton papers at Hurstbourne Park but had found other unpublished evidence suggesting that in 1692–93 Newton had had a full-blown nervous breakdown, which had left him incapacitated for some months and from which he had never fully recovered. The new documentary material was crucial to Biot's approach. Biot argued that Newton's life should be subject to the kind of objective scrutiny to which he himself had once subjected the natural world. The evidence came from a 1694 letter that had recently been given to Biot, written by Christiaan Huygens, a Dutch mathematician and experimentalist. He and Newton were acquaintances and had exchanged many letters. Huygens came to learn of Newton's "fall" into dementia through a third man, known only as Colin, who informed him that eighteen months earlier, the great man had "become deranged in his mind, either from too great application to his studies, or from excessive grief at having lost, by fire, his chemical laboratory and some papers." Colin went on to say that Newton had demonstrated the "alienation of his intellect" before the chancellor of Cambridge and that he was confined to his house and taken care of by his friends. The remedies had been successful enough, said Colin, that Newton could begin to again understand his own *Principia.*[4] Until then, according to Biot, Newton could not even understand his own greatest work.

Biot corroborated Huygens's letter with an extract from the diary of a fellow Cambridge don, Abraham de la Pryme, who also referred to a fire in Newton's Cambridge rooms that had evidently destroyed some of Newton's papers and equipment. In an entry dated February 3, 1692, de la Pryme noted that "when Mr Newton came from chapel, and had seen what was done [by the fire], every one thought he would have run mad, he was so troubled thereat, that he was not himself for a month after."[5] Biot suggested that Newton's breakdown was caused by the strain of losing these papers, as well as the intensely hard work that had gone into creating them. This could result in such distracted behavior as forgetting whether he had eaten or not and remaining in his dressing gown all day while at work on a problem. The many years of focused thought that Newton had devoted to natural philosophy, suggested Biot, were to blame. This work had taken him, like an extreme athlete, to the very limits of human ability. The passionate devotion and concentration that Newton exhibited, the defining characteristic of genius, could easily slide into mad obsession. It could also dissipate with astonishing speed. For Biot, a corollary of this kind of genius was that, as a flaring match leads soon to darkness, Newton's flame faded quickly. After the fire, Biot wrote, Newton "*never more*" produced "a *new* work in any branch of science," instead occupying himself with polishing pieces he had written previously.[6]

The vivid and inspired period of Newton's youth, suggested Biot, had given way to an insipid dotage during which he had done the majority (but not the totality) of his theological work, alongside the administrative duties he took on as master of the Mint and president of the Royal Society. The best Biot could say of Newton in this period was, as Huygens had written, that he recovered sufficiently to understand the contents of his own *Principia*, completed before the breakdown. Otherwise, Biot concluded, the breakdown had cleaved Newton's life in two incommensurable halves: the first contained the scientific breakthroughs, the second the theological obsessions. The implication was clear: the science was eminently rational, the religion decidedly irrational. From Biot's perspective, the breakdown served a useful purpose by keeping the stunning natural philosophy separate from the tainted theology.

Response to Biot's article in English was prompt and dramatic. The book was dismissed by a reviewer writing for the *Quarterly Review* as "a Frenchman's libel on the greatest of English philosophers, in which, *inter alia*, it is insinuated that his mental faculties had lost their vigour before he thought of writing on theological subjects."[7] For many, the question was whether Biot was driven by "enmity to England or to Christianity" to insult the memory of Newton. Either way, it was clear that Biot's article had thrown down a gauntlet to those for whom British honor was wedded to her greatest natural philosopher.

The publication of Biot's article coincided with another revelation that was equally alarming. It appeared in the same year in a new biography by Peter King of John Locke, with whom it was well-known that Newton had been friends. In his *Life of John Locke* King revealed an extraordinary letter that Newton had written to Locke, dated September 16, 1693, in which he apologized for extreme behavior that included having accused Locke of endeavoring to "embroil me with women" and to "sell me an office"—apparently accusing Locke of a kickback scheme—as well as apologizing for having thought of his friend "'twere better if you were dead." Locke's reply makes it clear that he had no idea what Newton is referring to, making Newton's letter all the more revealing. The date of the letter suggests that Newton was still unwell in September 1693, nearly a year after the Cambridge incident. Insomnia had something to do with it. Newton explained to Locke that part of the reason for these outbursts was that "when I wrote to you, I had not slept an hour a night for a fortnight together, and for five nights together not a wink."[8] Taken together, these documents make clear that Newton had indeed been mentally ill for a period of time during the 1690s.

While Biot's piece and King's biography caused a sensation, rather than give two publications too much of the credit for the surge of interest in Newton, we must widen our scope and turn to the "gentlemen of science" of the 1830s. We find among them plenty to account for the jolt of interest in Newton in the first third of the nineteenth century, on the cusp of the Age of Victoria (who ascended to the throne in 1837). These men, many of whom were instrumental in founding the British Association for the Advancement of Science (BAAS) in 1831, were, as their self-chosen title reminds us, still frequently gentlemen in the literal sense of having the property, the income, and the leisure to pursue science as an avocation rather than as a means to earn a living. Among them was Roderick Murchison, a wealthy Scot whose main distinction was in the hunting of foxes before he turned to the study of geology and established the first classification system for rocks. Others were less well-born, such as William Buckland, a vicar who was also president of the Royal Geological Society and in 1824 made the first full account of a fossil dinosaur, the Megalosaurus. Still more humble were the origins of William Whewell, a country lad from Lancaster who rose to towering prominence as master of Trinity College, Cambridge, thanks to notable contributions to, among other things, the study of tides, minerals, astronomy, mathematics, and physics, as well as the history and philosophy of science. The son of a teacher, David Brewster made many original investigations into the diffraction of light (and invented the kaleidoscope) while struggling to support his family on his cobbled-together earnings as a freelance science writer. Linked by a shared commitment to the elucidation of nature and its applications to practical improvement, Murchison, Buckland, Whewell, and Brewster belonged to a group that was big enough and diverse enough for the first time to feel the need for unifying institutions, procedures, even a name. In 1833 Whewell provided the latter at the third meeting of the BAAS, suggesting *scientist* as a convenient term for a group of men who cultivated science in general. "Thus we might say," Whewell explained, "that as an Artist is a Musician, Painter, or Poet, a Scientist is a Mathematician, Physicist, or Naturalist."[9]

The new revelations about Newton shocked many within the BAAS anxious to shore up their new, collective identity. Of these, Brewster devoted himself most wholeheartedly to defending Newton's good name.

Brewster was obsessed with Newton from early youth, when he had gazed intently at the memorial to the natural philosopher Colin Maclaurin in Greyfriars Church, Edinburgh, which read simply "Newtone Suadente" (Follower of Newton). (Maclaurin was the man who had reluctantly informed Robert Wodrow about Newton's "peculiar thoughts" on prophecy.) Brewster's admiration for Newton only grew. Later he visited Woolsthorpe and took away a graft from the famous apple tree. Brewster trained initially for a career in the Church, but an anxiety about public speaking—so acute that he once fainted while saying grace at a dinner party—made it painfully apparent that both sermonizing and academic lecturing were to be impossible for him. Instead he labored to support himself, his wife, and five children by writing and editing short articles and reviews. His was the harried life of a hack.

Constantly on the lookout for new projects to keep a steady stream of income flowing, Brewster was primed to understand the need for consistent financial support for scientific research. He found himself in the middle of growing discontent over the status of science in Britain. Traditionally individuals rather than the state were expected to fund their own investigations, but the Industrial Revolution, and anxiety about maintaining supremacy over France, meant that things were changing. Many men, Brewster among them, argued that the time had long passed when solitary tinkerers and experimenters could propel the advancement of science and its practical applications. Substantial public investment in science was necessary to secure a safe future. With

David Brewster was obsessed with Newton from an early age. Reliant upon his earnings as a freelance writer, he fought to establish the importance of science in Britain. A lithograph from the 1830s. © National Portrait Gallery, London.

no institutional affiliation other than the positions he held within the BAAS, Brewster became a central figure in the push to create a solid institutional basis for science in Britain. As the paragon of British science, Newton was a crucial resource in the fight for support from state and industry. When Biot's memoir appeared, Brewster felt that it was incumbent upon him to defend the spirit of Newton—and the spirit of British science itself.

Brewster decided that the best defense was a good offense and set about writing his own biography of Newton. His main concern was to respond directly to the accusation that Newton had gone mad. He considered it to be nothing less than a "sacred duty to the memory of that great man, to the feelings of his countrymen, and to the interests of Christianity itself"[10]—though he would be forced to acknowledge some of the other troubling revelations about Newton that were becoming increasingly public.

Though he too failed to convince the Portsmouth family to allow him to access the papers at Hurstbourne Park, Brewster did his best to uncover new material for the cause. Advertisements for the book boasted that its author had "sought out from resources, hitherto unknown, and inaccessible to previous writers, every fresh and novel particular regarding his life." This was strictly speaking true, although Brewster was looking only for evidence that would help him refute Biot's assertion. These new sources were assembled with the decidedly partisan aim of throwing "a great and agreeable light on the supposed insanity of Sir Isaac Newton."[11] Brewster worked fast, and by 1831 his *Life of Sir Isaac Newton* was available in an accessible duodecimo edition, costing just 5 shillings.[12]

But in the end Brewster hadn't found much. Worse, what he did find didn't enable him to claim unequivocally that a breakdown had not occurred. The evidence Biot had mustered was too strong. However, like any good prosecuting attorney, Brewster did cross-examine Biot's evidence to show up its weaknesses. Did the dates of the de la Pryme and Huygens reports match up adequately? (In fact there were puzzling discrepancies: de la Pryme's report referred to an event before January 1692, while Huygens's referred to one in November 1692.) Hadn't Newton written eminently coherent letters to another correspondent, Richard Bentley, the scholar and theologian with whom he had communicated about gravity and God, during the period when he was supposedly incapacitated? Finally, Brewster called some new witnesses onto the stand: a set of newly available correspondence between Newton, Samuel Pepys, and John Millington, a medical doctor, that demonstrated, Brewster argued, that Newton's illness was purely physical, the result of a lack of sleep and appetite and the added anxiety of his search for a new position in London.

Brewster admitted that a letter from Newton revealed that he had lost his "former consistency of mind" for as long as a year. And Pepys too had mentioned a letter from Newton that could point to "a discomposure in head, or mind, or both." But Millington had written a letter in reply to Pepys that soothed Brewster, just as it had no doubt been intended to soothe Pepys: "He is now very well, and, though I fear he is under some small degree of melancholy, yet I think there is no reason to suspect it hath at all touched his understanding."[13] Brewster rejected the idea that Newton's breakdown was anything more than a singular event and scoffed at the idea that it could have been precipitated from something so minor as the loss of some papers.

More significantly Brewster sought to integrate the natural philosophy and theology into a coherent body of rational work. To do that, it was important to show that Newton had in fact done theological work before the events of 1692–93 and scientific work afterward. After all, men of science in Britain were understood to be, in their highest manifestation, also men of devotion, and it was slander to suggest, as Biot had, that Newton's religious work was nothing more than the ravings of a diminished mind. But there was no going back to the simple pieties of the earlier century. The contemporary vogue for scientific history, based on extensive use of primary source materials and the objective presentation of evidence on which the reader was urged to make his or her own judgment, put Brewster on the defensive. The cat was out of the bag, and it was no longer possible simply to deny the existence of Newton's breakdown. The trick was to portray it as a temporary riffle on an otherwise calm mental sea.

One of the problems Brewster faced in that regard was the nature of Newton's genius. Instead of being born of diligence, his discoveries—or the stories about them—were shot through with images of startling clarity and sudden epiphany. The apple that fell from the tree at Woolsthorpe and triggered his thoughts on universal gravitation, a story of which Newton himself was the author, stood for all the ineffable quickness of his vision. So too did the piercing rays of light that he used in his crucial experiment to show that white light is actually made up of colors. As dramatic as they were, these stories risked casting Newton as a solitary, somewhat unhinged Romantic genius, visited by bursts of inexplicable inspiration. They were therefore quite unsuitable as pedagogical tools. Brewster's response was to distinguish between the less salubrious "poetic" genius—Samuel Taylor Coleridge during his drug-addled phase was a famous example—and what he defined as the more sturdy "philosophical" type, which was more likely to last. Just because Newton had flowered with an early "exuberance of invention," Brewster cautioned, he was not to be mistaken for a mad poet.[14] Brewster argued that science and faith had been linked throughout a long and pious life. There was no before and after to his genius, as Biot had claimed.

Brewster's biography of Newton sold well as a low-priced mass-market book, going into eight English editions and twice as many American ones and appearing in both German and French translations. Reviewers of the work agreed that Brewster's history of science was admirable (if, they added, somewhat dumbed down) but critiqued his lack of impartiality when it came to Newton. One reviewer, Benjamin Malkin, writing in the *Edinburgh Review*, criticized Brewster for letting his "enthusiastic admiration for his subject" get in the way of the truth. In particular Malkin felt that Brewster did "not give much credit to his judgment" about three topics in particular: Newton's alleged poverty (for some, the true mark of a Christian), his dealings with Leibniz, and his temporary insanity. It was clear, wrote Malkin, that Newton had lived in comfort for most of his life and that he had most certainly acted in vindictive and dishonest ways in relation to the dispute over whether he or Gottfried Leibniz could claim priority in the discovery of the calculus. It was also clear that he had indeed gone briefly but definitely mad.[15] The fact that new information on Newton could be uncovered more than a century after his death struck Malkin, for one, as highly significant and seemed to justify the interest in Newton's breakdown. Biot had done a creditable job, but he was writing about Newton's mental health "in a very different stage of the investigation," before the new information uncovered by Brewster and others had been available.[16] Thanks to the accumulated evidence, wrote Malkin, it was now undeniable that at one point Newton had indeed lost his mind.

Madness will out, and it had. Now the question was how to limit the damage. As the letters that had been published in King's biography of Locke had indicated, it wasn't just the contents of the Portsmouth papers or of competing biographies of Newton that had to be policed. Papers lay scattered in unknown dark corners. At any moment documents could emerge to further besmirch Newton's reputation. Following fast on the heels of Brewster's hopeful but ultimately ineffective defense, a new cache of material relating to John Flamsteed, the first Astronomer Royal and a famous antagonist of Newton's, emerged from a dusty garret to put into question once more the content of Newton's character.

And so, despite Brewster's best efforts to the contrary, the spirit of Newton would not be allowed to rest unmolested. Though there were no reports of the ghost of Isaac Newton appearing at the fashionable séances of the age, his spirit was to haunt the pages of several more provocative publications before the century was over.

5

# The Meanness of Newton

The 1833 meeting of the British Association for the Advancement of Science was a busy one. William Whewell coined a new word, *scientist,* to describe the assembled investigators. An ailing Samuel Taylor Coleridge turned up, lending an artistic air to the proceedings and querying the direction this rising generation was going. And the banker-turned–historian of science Francis Baily pointed to a pile of dusty papers on the table before him and made the startling claim that they revealed a truer portrait of Isaac Newton than had previously been available.[1]

Standing in Cambridge, the very town where Newton had made his great discoveries, Baily called for a reappraisal of the greatest founding father to which the fledgling Association could lay claim. We must go back to our origins, suggested Baily, and reappraise our myths if we are to move forward honestly, with our account books in order. The ledgers must be tallied, points for and against Newton rescored, rectified. The metaphors began piling up. Overlying it all was the legal language, by then a part of science, of proof and evidence. The facts must be revealed, said Baily with another gesture toward the papers that he had brought to the meeting, and the public left to make its own decisions. If this last suggestion was disingenuous (Baily believed that the facts revealed in the papers that lay on the table in Cambridge were overwhelmingly unflattering to Newton), it was increasingly important to uphold the pretense of objectivity in one's allegiances. The documents were well and truly coming into their own as tools in a struggle for the right to use truth to put forward one's case. And what was science, after all, but a codified method for assigning truth value to one's statements?

Francis Baily brought to light the Flamsteed papers and their evidence of a vindictive Newton. He was described by his friend John Herschel as having a personality "like a sphere whose form is perfect simply because nothing is protuberant." © National Portrait Gallery, London.

Baily may have been driven by partisan values, for or against Newton, and whatever he was seen to represent, but turning to the documents, and pointing at them, *that* was calculated to seem neutral, objective. It was a dramatic gesture calibrated to a world of newly surging print, of steam presses and photogravure, of weekly journals and daily newspapers. In a world of multiplying publications, the original document had become necessary and distinct in a way it had never been before.

What exactly was in the papers? The Earl of Portsmouth remained cagey about sharing his Newton trove too easily, and the papers that Baily put down on the table before the assembled members did not come from Hurstbourne Park. Instead they were letters addressed from the Astronomer Royal John Flamsteed to his friend and former assistant Abraham Sharp as well as letters from another of Flamsteed's assistant, Joseph Crosthwait, to Sharp. Many concerned the contentious publication of Flamsteed's magisterial work on star identification and location, *Historia Coelestis Britannica* (British Catalogue of Stars), which Newton had played a major role in bringing to light against Flamsteed's wishes.

While Brewster searched actively (if unsuccessfully) for evidence with which to defend Newton from Biot's accusation of madness, Baily's investigation of Flamsteed and his relations with Newton had been prompted by chance. A neighbor showed him a collection of letters from Flamsteed that were lying "in a garret filled with books and papers."[2] Baily reported that they had been originally found in a box deposited in an old lumber room, filled with miscellaneous papers that were being steadily depleted as the servants used them to light the fire. Luckily about 120 letters escaped. The correspondence, Baily reported, contained information of interest to any scientist committed to discovering the truth about such intellectual forebears as Flamsteed, Savilian Professor of Astronomy Edmond Halley (the man who had funded the publication of the first edition of the *Principia*), and, above all, Isaac Newton.

Baily's presentation was in many ways anomalous: he was the only speaker in the only section on history of science in a meeting dominated by presentations in the exact and natural sciences. In another sense, however, Baily's presentation was completely and even tellingly appropriate for a scientific meeting. In pointing to the table on which lay his papers, Baily was announcing as exciting a discovery as the other men presenting their findings at the meeting. If his discovery was a "mere" pile of paper, so too were the discoveries of other scientists, who presented reports based on experiments whose authenticity (and priority) was proved via notebooks and journal publications. Discovery was a problem these Victorian scientists grappled with strenuously, for discoveries were the paving stones upon which science progressed. But how to fully understand the nature of a discovery and, more to the point, the inspiration necessary to achieve one? Could rules be drawn up that would guide scientists to them?

Whewell had opened the meeting with a speech in which he addressed the question of what such an association could hope to accomplish. Each science required its own great men, geniuses like Newton, who could synthesize vast tracts of knowledge in order to generate mathematical predictions of how the world worked. Such men, according to Whewell, were born, not made. As yet astronomy was the only science to have achieved such a status, thanks in no small part to Newton, though other areas, such as optics, were surely not far behind.

Whewell cautioned his audience to remember the limits of what they could hope to achieve in their meeting. Let none suppose, he said, "that we believe in the omnipotence of a parliament of the scientific world." There is no such thing as a "royal road to knowledge" that may be confidently planned out from the safe and smug confines of the meeting hall. Science could not be willfully directed. Instead each man must start from his actual position. There was no way to "accelerate our advance by any method of giving to each man his mile of march."[3] No dividing and conquering, no organized parceling out of tasks could reduce the project of science to a mere tallying up of facts. For that, a genius and a sudden insight into the underlying order of the natural world were needed.

In pointing to the papers on the table—papers that bore on Newton's character—Baily was, like Whewell, drawing the attention of his fellow men of science to a central truth of their growing, specializing discipline: discoveries were essential and essentially serendipitous. They were also, in the fields of both history and science, archival. They could be made only retrospectively and on the basis of paper records. By presenting his material at the BAAS meeting, Baily made it clear that his methods—and his intentions—were as unimpeachable as those of the physical or natural sciences. He also made it clear that old bromides about scientific saints did not stack up against the evidence of the archive. David Brewster could not fail to have listened and taken note.

Baily's tale emphasized how the fate of *Historia Coelestis* hinged on the status of paperwork. The letters on the table revealed new facts about an old scientific controversy with Newton at its center. The story started in 1694, when Newton, seeking astronomical data to bolster his theory of the moon's orbit in preparation for a second edition of the *Principia,* first asked Flamsteed to share his observations with him. Flamsteed, a perfectionist who had taken on the huge task of creating a new chart of all visible stars, was wary of releasing any data that hadn't been thoroughly checked. He reluctantly agreed to share some of his lunar observations on the condition that Newton keep the material private. This cautious civility was not to last, as Newton, impatient and focused on publication, became increasingly eager to obtain the calculations that could help verify his theories and that only Flamsteed, with his vast trove of data, could provide. If the material was to be made public, reasoned Newton, then he could use it for his own purposes. In 1705 Flamsteed was forced to send a sealed manuscript containing his still incomplete star catalogue to a Royal Society committee of which Newton, not incidentally, was a member. At work behind the scenes, Newton saw to it that in 1711 the seal was broken and the incomplete *Catalogue* sent to press without Flamsteed's consent. Matters were made worse by the substitution of Flamsteed's own preface, condemning both Newton and Halley, for one that criticized him. Much aggrieved, Flamsteed demanded the return of the manuscripts, but the Royal Society committee (controlled by Newton) refused. Flamsteed's displeasure may be gauged by the fact that he managed to obtain and destroy three hundred of the four hundred printed copies of the unauthorized star chart. He never did get all of his manuscripts back, and he died a defeated man, having managed to publish just one volume of the stellar observations that had taken up nearly nineteen years of his life. (An "authorized" version of Flamsteed's *Historia Coelestis Britannica*, containing the position of more than three thousand stars, was published only in 1725, some six years after his death.)

At first glance Baily seems an odd type to enter into this old battle. While Brewster was a man perennially dissatisfied and prone to nervous anxiety, Baily was almost preternaturally calm and self-assured. The astronomer John Herschel struggled to find words to express the nature of Baily's personality, which consisted of a kind of extreme blandness, as unobjectionable, as he put it, as "a sphere whose form is perfect simply because nothing is protuberant."[4] But enter into the fray Baily did, serving as a proxy for Flamsteed as he declaimed the values of objectivity.

Baily's mild personality belied an early life filled with adventure. Born in Berkshire to a successful banker father, he began a City apprenticeship at fourteen, with the understanding that he would enter the family business. But at seventeen he discovered in the writings of Joseph Priestley, a radical dissenter and chemist who discovered oxygen, a sympathetic union of scientific rationalism with democratic intellectual and political ideals. In his early twenties, Baily's passion for democracy led him to America. He toured the country, championing the rights of slaves and the American ideals of individualism, liberty, and free commerce. Traveling alone through the untamed American wilderness, he camped out for nearly a year in Indian Territory and survived two shipwrecks and floating in an open boat from Pittsburgh to New Orleans.

After these swashbuckling adventures, Baily's life took an unexpected turn back to normality and to England, where, failing to secure further funding for a planned trip to Africa, he dutifully followed in his father's footsteps. Not content merely to follow established tradition, however, Baily became a pioneer in the field of life insurance, publishing the actuarial tables of estimated lifetimes that were helping to establish the nascent practice of insuring lives. He then spent more than twenty-five years as a successful banker.

Baily's retirement from banking was anything but restful, filled instead with energetic and quantitatively rigorous feats. His first task was to inaugurate what he hoped would be a new age of truly rational history, illuminated by his "New Chart of History." Completed in 1812, his "New Chart" did for past human endeavor what his mortality tables promised to do for future human lives: reduce them to predictable, orderly tables susceptible to calculation and analysis. If the length of a human life could be reduced to a table of probabilities, why should not all of history be susceptible to the same method?

Baily's newfound interest in history led him directly to astronomy, the perfect historical tool for a man possessed of a passion for calculation and a drive for precision. The subject of Baily's first paper to be read before the Royal Society gives a flavor of his approach. The paper attempted to fix a date for a celebrated total solar eclipse, said to have been predicted by Thales, and described by Herodotus. While it was theoretically possible to extrapolate backward from lunar tables to determine the date for the event Herodotus described, in practice lunar tables were not accurate enough to allow it. Baily set out to calculate all solar eclipses that had occurred during a nearly century-long period six centuries before the Christian era. He followed this feat of astronomical history with his "Epitome of Universal History," which accompanied a "historical chart" that represented the whole of history in graphical form.

With his twin interests in history and astronomy, Baily quickly established himself as a leader in the astronomical community. He helped to found the Royal Astronomical Society in 1820 and set himself the unenviable task of revising historical star charts created by men such as Ptolemy, Tycho Brahe, Halley, and Hevelius. These efforts required a nearly Herculean supply of attention to detail and an almost evangelical belief in the importance of linking past observations with present and future ones in order to create an unbroken chain of authenticated, accurate data from which both historical and scientific truths could be attained. These projects in historical reconstruction were exceptionally good preparation for the opportunity that followed the chance discovery in his neighbor's garret.

Baily was nothing if not persistent, and, having found such rich material so close to home, he decided to dig further. A search at the Royal Observatory struck gold. There he found, "to his great surprise and delight," Flamsteed's papers, which had been sitting on the shelves of the library for sixty years, "unnoticed and unknown." Among them Baily discovered Sharp's replies to Flamsteed, Flamsteed's own original entries of his astronomical observations made at the Royal Observatory, numerous pages of his calculations on these observations, Flamsteed's suppressed preface, and what Baily could only describe as a "vast collection of letters" written between Flamsteed and correspondents at home and abroad.[5] Just as the fossils that were being unearthed on railway cuttings across the nation challenged established ideas about the age of the Earth and the creation, and extinction, of God's creatures, these manuscripts had been lying just under the surface, waiting for the time when their true significance would be grasped. With Baily's keen eye upon them, their time had come. And none too soon. The papers were in "great confusion and disorder." Most books had lost their covers, letters and papers were scattered everywhere, and, worst of all, so much paste had been used to fasten them into so-called guard-books that they were "literally mouldering away."[6] Baily performed some basic first aid on the documents, separating them from the paste, sorting them, and rebinding them.

And he continued to hunt. He penetrated archives in the British Museum and libraries at Oxford and Trinity College, Cambridge, and finally managed to do what Brewster had not: persuade the Portsmouth family to give him access. In April 1835 he traveled out to Hurstbourne Park to have a look at the trove of Newton papers himself. He appreciated, as Horsley had not, what treasures the papers contained. Noting that the general assumption that they were of "little or no value" was based on the judgment of Pellet immediately following Newton's death, Baily pointed out that Pellet had been charged with assessing the papers' financial prospects. Baily set this record straight too: though the documents may not have been likely to raise money for Newton's heirs, they undoubtedly contained much that was new and illuminating. He hoped that they would soon be examined for the purpose of cataloguing and then publishing some portion of them to "illustrate more fully (as I am sure they will do) the life and labours of our illustrious countryman."[7]

Baily had done a remarkably thorough job of tracking down historical documents, many of which bore on Newton's personality and treatment of others. The "illustration" of Newton's life that they furnished was not a pretty one. The papers revealed Newton to be an angry, vindictive, and even duplicitous man who used all available means to pursue his own goals, whatever the cost to those who opposed him. In one journal entry, Flamsteed recounted how an enraged Newton had called him "many hard names; *puppy* was the most innocent of them."[8] On Newton's having shared the lunar observations that he had promised to keep private, Flamsteed remarked that Newton was a "lazy and malicious thief."[9] Newton's duplicity extended further, to the unforgivable lie that Flamsteed's sealed catalogue had been opened under order by the queen, a blatant untruth. Newton, Flamsteed claimed, was "no friend" to his work. Rather his "design was either to gain the honour of all my pains to himself, to make me come under him . . . or so spoil and sink it."[10]

Flamsteed, in contrast, appeared "in a light very different from that in which he has been generally viewed." Because he had taken so long in preparing his observations for the press, many had accused Flamsteed, during his lifetime and in the hundred years since, of being a "selfish and indolent observer" who had hoarded data that might have aided maritime navigators and thus saved lives at sea. Baily's documents showed how very active Flamsteed had been—making and paying for his own instruments and devoting all of his spare time to pursuing lunar and planetary theories, trying to explain anomalies, and, perhaps best of all to the actuarial Baily, "forming tables for the more accurate computation of their places, and communicating the result of his inquiries with the greatest readiness to those who were prosecuting the same studies."[11] In other words, Baily decided that Flamsteed looked a lot like him: a patient observer and tabulator. Baily had found the necessary ammunition for a full-scale defense and resuscitation of a man with whom he could not fail to identify.

In order to set the record straight, and to minimize the risk to his own reputation in publishing on such a contentious topic, Baily's lengthy "Account of the Revd. John Flamsteed, the First Astronomer Royal," to which he added Flamsteed's "British Catalogue of Stars, Corrected and Enlarged," published in 1835, consisted almost entirely of historical documents. Baily had taken a novel and, in many ways, prescient approach to formatting his book. The work of more than eight hundred pages was divided into two parts. The first contained an extensive autobiographical memoir by Flamsteed himself, a catalogue of Flamsteed's manuscripts, and an appendix including almost three hundred pages of correspondence between the astronomer and prominent contemporaries such as Christopher Wren, Halley and, Newton.

The second part of the book consisted entirely of Baily's three-hundred-page fastidiously corrected version of Flamsteed's original *Historia Coelestis Britannica.* By publishing a book of two halves, in which the written documents were made the equal and opposite partner to a set of putatively bias-free numerical observations (the corrected catalogue of stars), Baily was demonstrating that history itself could be made into a science—objective, dispassionate, and value-free. The truth, claimed Baily, was the highest object, and the documents should speak for themselves.

And what they said was not flattering to Newton. Baily presented himself as a neutral seeker who had searched for corresponding documents "of a contrary tendency" that might exonerate Newton and Halley.[12] He had accessed the trove of papers at Hurstbourne Park, looking for more evidence with which to try the case. But in the end he found nothing that changed the verdict that Newton had acted in a two-faced and vengeful manner, wielding his considerable influence to serve his own ends. If Biot had been eager to expose Newton's madness, and Brewster to save Newton's piety and sanity, Baily showed that Newton had been, when it suited him, a very mean man.

Why, Baily asked in his preface, had these valuable documents been hidden away for so long? The only reason he could come up with was that previous editors of a life of Flamsteed "might not have thought it prudent or politic to risk" publishing documents that reflected so negatively on the characters of Newton and Halley.[13] Baily acknowledged the discomfort that readers might feel in learning that the famous men had been "mixed up" in such unsavory business but stressed that "a proper regard for truth and justice" required the publication of the facts contained in the manuscripts, however lamentable they might be.

That phrase "a proper regard for truth and justice" was the rallying cry for a very personal campaign, once fought between Flamsteed and Newton themselves, now bringing Baily and Brewster in a head-to-head battle, pitting against each other two portrayals of Newton and two assessments of his legacy. Baily's *Account* formed a devastating denunciation of Newton (and Halley, among others). The task for the burgeoning scientific community was not simply to evaluate the enormous set of new documents that Baily had made public but to continue to seek out more. Baily had presented a potential new discovery about Newton's meanness. It was left to other men to put his discovery to the test.

Despite Baily's assertion that the public should judge the facts about Newton for themselves, his *Account* was not available for sale to the general public but instead was distributed directly to a highly select group of individuals and scientific academies, observatories, and universities in Britain, Europe, and America. Thanks to this careful marketing, the 250 copies that were published succeeded in inflaming the passions of those who sought to preserve Newton's memory unsullied. The Admiralty had paid for the expensive work to be printed in order to present the corrected star catalogue—essential for British mariners seeking to plot a course—in the best light. But it was the biographical, and highly personal, material in Baily's *Account* that inevitably generated the most attention.

Foremost among those who found ample material to which to object was Whewell, a man of towering importance in the history of

William Whewell believed that science brought its practitioners close to the mind of God and that Newton had gotten the closest of all. © National Portrait Gallery, London.

science of the period who is today almost entirely unknown. There is something fitting in his lack of prominence today. Whewell, who argued that discoveries alone mattered in science, himself made none. Instead he sought to make a role for himself as a commenter on and "critic" of science. What fame he has today rests in the field of history and philosophy of science.

Whewell was born to a master carpenter in Lancashire. After excelling in his local school, he entered Trinity College, Cambridge, thus traveling directly from the countryside to the university, with no intervening experience of the machine culture that so profoundly reshaped the world in which he lived.[14] He remained at the university his whole life, becoming perhaps the only person ever to transition effortlessly from a chair in mineralogy to one in moral philosophy.

It was as a moral philosopher that Whewell responded to Baily's *Account* and the accompanying new evidence on Newton that had been published by Brewster, Biot, and others in previous years. Whewell saw much to lament in the radical changes that were reshaping Britain in the 1830s—not least the exponential growth of the railways across a once pastoral England. He was profoundly concerned with how the progress of science and technology affected the moral character of its practitioners and vice versa. He argued that the best form of thinking—and the only one that yielded true discoveries, that zenith of scientific endeavor—depended not on automation and systematized rules (as did so much of the new machine culture) but on individuals who possessed the correct moral demeanor. As he argued at the British Association meeting of 1833, the former outsider, now a Cambridge don, strongly believed that science relied upon an elite group of insiders—call them geniuses—who had the power to see the underlying order through the mass of accumulated details that made up scientific data. Such a power was in itself nearly divine. Making scientific discoveries meant studying the very intelligence of God, a study that could not fail to inspire strong faith. "When they had just read a sentence of the table of the laws of the universe," reasoned Whewell, the true discoverers of science "could not doubt whether it had had a legislator."[15]

Whewell was therefore primed for something as incendiary as Baily's *Account* to come along. He found in it much to dislike, not least Baily's irksome tendency to exalt the most mundane and banal of scientific activity—mere observation—while critiquing the conduct of the greatest scientist Britain had ever produced. He seized upon Baily's book for the opportunity to argue on behalf of a particular form of thinking. There was no one better than Newton to demonstrate the value of pure thought. Whewell responded with a pamphlet titled "Flamsteed and Newton" that aimed not just to save Newton's reputation but to settle the problematic relationship between the natural philosopher (or "theorist," as Whewell sometimes wrote) and the mere observer. Whewell distinguished between Flamsteed's observations and Newton's interpretation of them by saying that it was the difference between "the discovery of *what* occurred, and the discovery of *why* it had occurred—between an observer and a philosopher."[16] This was partisan stuff, and Whewell admitted as much. In writing the pamphlet, he identified himself with those who felt the same strong interest "in the fair fame of Newton which those cannot fail to feel who love to contemplate the union of intellectual and moral excellence."[17]

Whewell also emphasized that rather than opposing each other, Flamsteed and Newton had needed each other. Flamsteed's observations had the power to confirm Newton's theory of the moon as well as the more general theory of gravitational attraction. Newton himself implied as much in a letter of February 16, 1695, in which he reminded Flamsteed, "All the world knows that I make no observations myself." And Flamsteed needed Newton's theory because the congruence between it and Flamsteed's observations would be, as Newton put it, "a demonstration of their exactness" and show Flamsteed's consummate skill as an observer.[18] Nonetheless, while there was a mutual reliance between Newton and Flamsteed, for Whewell Newton's superiority was absolutely clear. Flamsteed ultimately depended on Newton because the theory was "so very intricate" that only someone who understood the theory of gravity as well or better than he himself did could perfect it. The person who understood the theory of gravity better than Newton, it was unnecessary to add, was unlikely to be found.

Whewell set out to win the argument, but there was no real victory to be had. As the documentary evidence mounted, reviewers began to accept that contradictory forces might be present within even the greatest men (and possibly especially so). For those able to detach themselves from partisan perspectives on the decline of science and Christian morals, there was room for a more capacious understanding of the nature of man. In a review of Biot's and Brewster's pieces in the *Foreign Quarterly Review* in 1833, the Scottish mathematician Thomas Galloway summarized the elements of Newton's character that the new documentary evidence (which did not yet include Baily's discoveries) revealed: "a sombre and retiring disposition," a certain amount of "timidity and suspicion," the "usual concomitants of a hypochondriacal temperament," an aversion to publication and controversy, and "a morbid constitution of mind," of the sort that often afflicts "the most highly gifted" and that made Newton susceptible to the "desolating ravages" of mental illness.[19] Galloway quoted approvingly from Biot's consideration of Newton's insanity: "Such is the frightful condition of man. Genius and madness may exist in his mind side by side and simultaneously."[20] This insight into the dual nature of Newton's character marked an important shift in the way it was becoming acceptable to speak about Newton. New facts gleaned from new sources were to be welcomed and assessed for their contribution to a fuller portrait of the man. The time had passed when a singular perspective would suffice to capture Newton.

Any hope of a more complete picture of the man meant looking at all the evidence. Following the success of his *Account*, Baily himself pushed for publication of the Portsmouth papers. Though he claimed that it was truth alone he was after, Newton's descendants remained unconvinced. When Baily managed to gain access to the papers it was Henry Fellowes, a nephew of the Third Earl of Portsmouth, who had showed him around. After his *Account* and a subsequent *Supplement* were published, it was Newton Fellowes, Henry's father (later the fourth earl), who responded to Baily's request in an icy tone. "I have always felt a jealousy with respect to the manuscripts being seen & in any way used without the full authority & concurrence of those whose property they are." The fourth earl went on to say that while he was open to the idea that "useful matter" might be made public, he would reserve the right to decide what that meant.[21]

In a further, 1836 review of Baily's book, Galloway confirmed that there were plans afoot to publish some of the papers, with rumors that the government was prepared to print them at public expense. Apparently the family was divided on the matter, and Galloway took the opportunity to urge them on. Publishing them now would be the "antidote to any unfavourable impressions." To Galloway's mind, Newton's character was a "species of national property"—indeed "the nation's glory"—and therefore deserved to be better known. Heirs could be sticky and recalcitrant, and Galloway may have sensed that the battle was not won. He hoped at least that the letters might be put "into a better state of order and arrangement than they are reported to be at present, and preserved with religious care in the anticipation of a more auspicious period."[22]

Following this outpouring of controversial material, Brewster devoted himself once more to the forensic excavation of his hero, a task he undertook only because he believed it offered the best, and perhaps the only hope of saving Newton's reputation once and for all. Just two years after the publication of Baily's book, Brewster seemed to have achieved a real breakthrough in this biographical game of cat and mouse. In May 1837 Brewster finally achieved what had seemed impossible: he got permission to examine the Portsmouth papers himself and began a week of "intense labour" at Hurstbourne Park just two years after Baily's own visit.[23] Henry Fellowes helped him "night and day in Copying from the MSS," and together they went over all the papers "*twice*."[24] Brewster was faced with what we now know to be an overwhelming volume of uncatalogued papers, ranging on subjects from theology, chronology, and alchemy to natural philosophy, most of which demanded considerable expertise in their respective fields He nonetheless made a typically heroic stab at surveying the lot. He shared his delight in the sheer quantity of the material and its descriptive richness in letters to friends, mentioning that he had found materials "which were not supposed to exist" and highlighting what he called a "love letter" from Newton to Lady Norris, a widowed acquaintance.[25]

Faced with the onslaught of unfavorable material made public by Baily, Brewster set out to play his part in the game. If he could find documents that showed Newton to be generous, mild-mannered, and hard-working, he would have saved a great man's reputation for the nation. He planned to publish as soon as possible, but it was not to be. Ironically Brewster's habitual need to earn a living prevented him for years from finding the time to complete the work he hoped would vindicate Newton—and further defend the role of British science.

6

# Getting to Know the Knowers

By the end of the 1830s Newton had become subject to the same energetic, self-searching spirit that stimulated so much Victorian progress. At the very same time that Charles Darwin, voyaging on the *Beagle* from 1831 to 1836, was learning to place his own observations into the crucible of cross-examination, observers of Newton were increasingly aware that evidence alone could not generate increased certainty. For many, the future seemed to promise only new kinds of questions and growing alienation from what had once seemed certain and sacred.

Brewster's biographical retort to Baily and Biot languished on his desk for nearly two decades, victim to the busy demands of his work in reforming the University of St. Andrews (where he had finally, in 1838, secured a permanent post), helping to form the Free Church of Scotland, completing more optical research, and writing hundreds of review articles for the popular press. During these years attitudes toward time itself were changing. In geology and natural history, theorists proposed incremental and staggeringly slow processes of change. On the other hand, the railways, and increasingly standard ways of working according to the clock, made time seem to move faster and in more discrete segments. The social changes wrought by the new ways of working and living—crowded together in cities—also seemed to speed up things. Time came to seem strangely variable, newly fantastic in its speed and its extent.

In 1838, just a few years after Baily published his *Account*, Parliament passed an act establishing a Public Record Office. Records once scattered in fifty-six separate repositories throughout the country were brought together under one roof in London. No longer were the staff paid out of revenues from fees charged for searches but were granted salaries that freed them to work more independently. History was starting to become a profession, with a burgeoning sense of itself as an independent pursuit with a philosophical rather than merely technical or administrative function. The fascination with history extended to the present, where a new historical self-consciousness emerged. As a token of this new temporal awareness a novel phrase, the "spirit of the age," emerged among a group of writers, including John Stuart Mill, William Hazlitt, and Thomas Carlyle. And so the debate over Newton mattered not simply to the small group of men who concerned themselves with the history of science (though it did matter most to them) but to a wider culture freshly aware of the role of history in the present, of the present itself as a kind of history.

Aside from the rooms of the government record office, there were other places where old books and papers were receiving attention the likes of which they had never seen before. Individuals possessed of equal parts passion and cash conspired to create, entirely independent of government attention or support, a frenzy of interest in the written remains of a previous age. These weren't fossils but rare books and manuscripts, more plentiful and easier to come by. The consummate book collector Richard Heber managed to accumulate more than 100,000 volumes for his collection of early English poetry and drama. He spent more than £100,000 to do so and famously remarked, "No gentleman can be without three copies of a book, one for show, one for use, and one for borrowers."[1]

One guide to this heady world is Thomas Frognall Dibdin. At once the primary promoter, diagnostician, and sufferer of a peculiar malady he called "bibliomania," Dibdin provided the commentary and a good bit of the hot air for a bubble of speculation in the market for rare books and manuscripts that reached giddy heights in the first three decades of the nineteenth century. In 1809 Dibdin published a book that identified "book-madness" as a "fatal disease" among a disparate group of people: rare book dealers, the moneyed (mainly aristocratic) collectors who could afford their wares, and the cash-strapped men of letters who could afford only to witness, not participate in, the spectacles of the day. By popularizing the notion of a book-madness, the second edition of Dibdin's book helped to create one. Specifically it fostered the event that came to define the mania: the 1812 sale of a collection of exceedingly rare early printed books from the great library of the Roxburghe family. The event remains legendary in annals of book collecting for prompting a bidding war that took place between two great aristocratic book collectors, Lord Spencer and the Marquis of Blandford. Competing against each other and subject to the acquisitive passions brought on by their mania, together they drove the price of the jewel of the collection, a 1471 edition of Boccaccio's *Decamerone,* to an eye-watering £2,260 (£75,000 today).[2] Depending on how you look at it, the Marquis of Blandford, who ended up with the book, was either the winner or the loser in this battle.

The Roxburghe sale quickly saw the launch of a sort of Roxburghe franchise (and, as a secondary effect, the creation of a robust market for rare early editions). The sale was celebrated with a dinner whose attendees formed the roster of an eponymous club launched that evening. Membership (which peaked at thirty-one) was secured with a promise to reprint a rare volume of English literature, often of dubious literary merit. It was later noted that of these rare editions it could often fairly be said that "when they were unique there was already one copy too many in existence."[3] But that was missing the point: uniqueness was king. Literary taste, for those in the most extreme throes of the book disease, was fittingly sacrificed on the altar of pure collectability. Form became everything, content almost nothing. The Roxburghe Club was the first example of what would become by mid-century a widespread, even middlebrow passion for the literary past. In its more accessible form, the craze for reprinting or collecting early editions was not as extreme as the Roxburghe affliction but was more broadly significant. Books were more than mere tokens of prestige; they were emblems of an imagined past.

As industrialization and the vigorous forms of capitalism that fostered it grew more jolting in the 1820s and 1830s, nostalgia for a putatively shared past became increasingly important—a kind of balm for the bruises of the new machine culture. This nostalgia took many forms. History and archaeology became almost requisite pursuits for a certain type of genteel person. Every county in England boasted at least one antiquarian society dedicated to local history and curiosities. Further along the spectrum, specialized printing societies sprang up to publish documents—until recently, neglected and often inaccessible—relating to every facet of the history of England. The Surtees Society, founded in 1834, for example, proposed to dedicate itself to the publication of manuscripts that were "illustrative of the intellectual, the moral, the religious, and the social condition of [these] parts of England and Scotland" that once had constituted the "Antient Kingdom of Northumberland."[4]

The search for Britain's moral, religions, and social underpinnings could engage a surprisingly broad assortment of people. The market for rare books, and the corresponding interest in the history of Britain, brought together middle-class antiquarians, aristocratic collectors, and jobbing booksellers. The thrill of connoisseurship, the glee of the canny purchase or the merely demonstrative one was made respectable, even honorable, by the shared sense that the private libraries and estates from which these books emerged and were repositioned were themselves part of the patrimony of Britain. What was private became, by a rhetorical and symbolic sleight of hand, communal. As the author of an exhaustive treatise on the history of British libraries declaimed, "Some of the best possessions of a house and lineage, like that of the Spencers are, for use and profit, common possessions."[5]

No one was advocating that the doors of the great houses and libraries of the noble families be thrown open to the masses. Instead, by conserving intact the literary patrimony of the nation, such libraries were akin to the great country estates, which had been husbanded by aristocratic landowners for centuries. They acted as a source, however remote and obscure, for the knowledge that was being diffused throughout Britain at the time in the form of mass-market books. "The common books which pass into the hands of almost the humblest owe something of their merit," was the conclusion of the author mentioned above, "to the heaping up of rare and costly books in such collections as the Spencer Library—created with a liberal hand, and imparted with a liberal heart."[6]

Due to this process of diffusion, Britain made public some of its treasures without claiming them for the state. While the possessions of the French aristocracy had been forcibly seized during the Revolution, in Britain the process was piecemeal, even fudged. It wasn't close to systematic, but it was a shifting of what was considered out of the reach of the appreciative (in both senses of the word) eye of the crowd to something to which a value could be assigned. There was to be no more cozy shutting away of treasures. The aristocracy would be made to submit to the marketplace as surely as were the masses. But the market was an exotic beast: seductive, dangerous, and irrational. Prices fluctuated wildly. Only seven years after the great Roxburghe sale, Lord Spencer, who had been prepared to pay more than £2,000 for it, acquired that rare edition of Boccaccio's *Decamerone* for just £750.[7]

Manuscripts entered the auction scene in the early 1830s. They had been valuable before, but only if they existed as unique documents of a given text. Previously, once a published text appeared, the manuscript had generally been considered a worthless remnant of the composition process. All that was changing, as the steam-driven printing press threatened to cover everything in a veneer of print. Handwriting acquired a newly elevated status as print bloomed. Initially there was much confusion about what the value of such a thing was. In 1831 Evans of Pall Mall presented at auction a large group that consisted of the manuscript drafts of thirteen of the published *Waverley* novels of Sir Walter Scott. Here were pages that might previously have been considered "loose and foul," working documents to be discarded upon publication of a finished piece. Now they had come to hold value as records of the emotional flavor of a work's creation and of the process of composition.

Manuscripts promised to reveal the secrets of composition and, by extension, the nature of creative genius itself. The forms in which literary texts could appear were more numerous than ever before. Rare and prestigious first editions of the sort bibliomaniacs craved, expensive reprints professionals might afford, less valuable copes that circulated from the library, and cheap abridgments working men and women could purchase: they all contained the same basic material. Scott's novels, set in historical times (as in the popular *Waverly* series, which dramatized the legends of Ivanhoe and Rob Roy), were the first to find what we would call a mass market, being published in many formats and many editions priced for readers of different social classes. It is no accident that his manuscripts were also the first to find a market among collectors, albeit a much smaller one. The elite market for manuscripts—and the regard for genius it evinced—became possible only when print publications proliferated in ways that complicated the idea of just what a text was. And it is also no coincidence that Scott's novels, and the manuscripts that recorded their composition, were themselves historical, narrating stories from Britain's legendary past.

There was also increasing attraction paid to manuscripts as charged emblems of genius itself. That passion, the devotional kind, haunts this phenomenon from the start. The bibliomaniacs, and later autograph hunters, were as unapologetic about their desire to touch the faded hem of a master's genius as any pilgrim. They had taken the new genius of the Romantic movement on faith, as it were, and sought to fill a need that hadn't been present before: to get close to the men who best instantiated the form. Passionate, inspired, both awestruck and awe-inspiring, these men were creative in a sense that had only just been conceived. Samuel Taylor Coleridge and John Keats exemplified the form. They were poets, unique and inimitable, even if such an avocation were not to be recommended, given the evident risks that attended their outpourings. Coleridge famously used opium to provoke the reveries on which his poetry was based, while Keats burned with almost incandescent creativity until he succumbed to tuberculosis at age twenty-five.

It was hard to value such emotional emblems. At the Evans auction of Scott manuscripts in 1831, Dawson Turner, an experienced buyer and collector of rare books and manuscripts, offered just over £5 for the lot, prompting Dibdin to cavil loudly against the low prices brought by these manuscripts, which he believed were worth something closer to £1,000. "Where were ye, ye pains-taking, fiddle-faddling, indefatigable collectors of Franks—ye threaders of autographic scraps—ye Album-ites," when it came time to bid?" he sputtered in rage. "One would have thought that the "original drafts of these master-pieces of human wit, eloquence, and passion" would be valued more highly.[8] In the event, the thirteen lots brought in a considerable £317, roughly $25,000 today.[9] While there may not have been much method in the buying and selling of books and manuscripts in this period, there was much madness, a mania that seems, in retrospect, to mimic the fevers of the geniuses whose works inspired such frenzy.

The madness of this sort of collecting was not lost on its devotees, but this did not dampen its allure. Indeed the mania surrounding books and manuscripts in the period contributed to an atmosphere in which the very men appointed to safeguard Britain's finest collections of books and manuscripts were also sometimes responsible for their desecration, forgery, and theft. Bibliomaniac crimes, difficult to prove or disprove, were perpetrated by the leading collectors and bibliographers of the period. In some cases the lure of cashing in on the valuable books in their care was simply too great. But more frequently what motivated these individuals to cross the line was far less simple than mere financial reward. This included the collector William Upcott, a longtime librarian at the London Institution, who was accused of removing from a great house a mass of manuscripts that he had been hired to catalogue in 1813. The tables and chairs at his crowded cottage groaned under loads of books, portraits, autographs, and newspaper cuttings, some of them evidently not legally obtained. He was seemingly powerless to resist the lure of old paper, admitting that the very word *autograph* was as "great a cordial to my heart as is a glass of full proof Geneva [gin] to a Billingsgate fishwoman."[10]

James Orchard Halliwell lived this dual life to its utmost and played a central role in the establishment of the first society devoted to the history of science in Britain. He was accused of outright theft and yet remains celebrated to this day as a pioneer in his field. Halliwell took up his studies as an undergraduate at Trinity College in 1837, at the height of the furor over Newton and Flamsteed. The son of a wealthy businessman, he possessed an energy peculiar to the age. By the age of eighteen he was already the author of a series of nearly a score of biographies of mathematicians and scientists for the *Parthenon*; three articles on scientific literature for the *Magazine of Popular Science*; a pamphlet on the life of Samuel Morland, master of mechanics to Charles II; and a collection of manuscript fragments and treatises in medieval arithmetic and geometry titled *Rara Mathematica*, which was issued in parts beginning in 1838.

In a review of these last two efforts by Halliwell, an anonymous writer for the *London and Edinburgh Philosophical Review* praised the "intrinsic value" of the effort at scientific bibliography and went on to offer a précis of the state of scientific history in the country, as compared with literary and antiquarian excavations. By 1838 the problem was well-aired. "The little attention paid by Englishmen to the history of science in England is not a new subject of reproach. Almost every human pursuit has had its history investigated, its fragments published, and its cultivators biographied, except science," noted the reviewer. He offered an explanation, pointing out that the history of science involves a rare combination of knowledge and grace, and there was little real "fraternity between the taste for decyphering ancient manuscripts, and that for pure scientific investigation."[11] That, in a nutshell, was the problem that had bedeviled and would continue to bedevil the potential editors of arcane scientific material. Science and the "decypherment" of ancient manuscripts fit together awkwardly, if at all.

At Cambridge, Halliwell quickly transferred from Trinity College to nearby Jesus College. His biographer assures us that while it isn't clear exactly what he was up to during his few years at Cambridge, "it is safe to assume that he did not attend many lectures."[12] Instead he threw himself into the maelstrom of antiquarian scholarship then afoot, cataloguing manuscripts in libraries in Cambridge and probably the Royal Society, coediting an edition of "Scraps from Ancient Manuscripts," and editing alone a Latin history of Jesus College, as well as a 1725 edition of John Mandeville's fabulistic travel book, first available in the late fourteenth century. He helped to found the Cambridge Antiquarian Society and the Percy Society, dedicated to the publication of rare poems and songs, and increased his own collection of early manuscripts in the history of mathematics and astronomy.

That collection was to be an early victim of his penchant for overspending, and by 1839 he was already forced to consider a fire sale of his books and manuscripts, citing "misfortune—dire and unexpected & enough to crush me."[13] Halliwell tried to interest the bibliomaniac Thomas Phillipps in his collection, offering it for £250—less, he claimed, than he had paid for it. Phillipps was also in debt, and so the Halliwell collection of more than 136 scientific manuscripts, many of them medieval, and including the papers of the astronomer James Ferguson, papers on spherical trigonometry by John Collins (a correspondent of Newton), and autograph manuscripts from Pierre Gassendi and John Flamsteed, ended up with a bookseller who sold them to a buyer who donated them to the British Museum. Halliwell recovered from this setback remarkably quickly. Between 1839 and 1844 he edited four editions for the Camden Society (a printing society in the vein of the Surtees), began to prepare a history of early English manuscripts, and found time to write *A Few Hints to Novices in Manuscript Literature.* Barely out of his teens, he soon was a council member of the Camden Society. He also founded the authoritative-sounding, but disappointingly short-lived Historical Society of Science.

Begun in July 1840, the Society lasted all of six years. Halliwell was both secretary and treasurer, and the membership boasted the Duke of Sussex as its president as well as five fellows of the Royal Society. The twelve-member council was made up of two rough halves: men who belonged to the Royal Society and those who belonged to the Society of Antiquaries, with a few belonging to both. Since membership in the Royal Society in this period was no guarantee of scientific achievement, the Historical Society also sought out bona fide men of science. Of those they managed to land, Augustus De Morgan, professor of mathematics at University College London, and the physicist and theologian Baden Powell, Savilian Professor of Geometry at Oxford, were the most prominent. Annual subscription to the publications of the Historical Society (and thus membership) cost a pound. Francis Baily was one of the first to sign up, along with Michael Faraday, the discoverer of electromagnetic induction. By 1841 the Historical Society had an impressive roll call of 179 members, including forty-five fellows of the Royal Society, thirty-eight fellows of the Society of Antiquaries, and thirty-four members with standing in other learned societies. Thomas Phillipps was among the members, alongside Beriah Botfield, another book collector whose library at Longleat House, the Elizabethan-era stately home in Wiltshire, was a prime example of the great noble collections. Foreign honorary members were also elected, including the French mathematician Michel Chasles and the Italian mathematical physicist Guglielmo Libri. By June 1840, when Halliwell invited Whewell to join, there were many societies soliciting the limited pool of Royal Society fellows and other eminences for their subscription fees and their time. Whewell, who had known Halliwell at Trinity, responded to the invitation by saying that though he admired the Society's "objects," he already felt be belonged to too many associations.[14] In his polite refusal is the first hint of the Historical Society's impending failure.

The Historical Society of Science was a cognate of the other printing societies, founded on the Roxburghe example, for which literary merit, as we've seen, was a secondary consideration. The restoration of past editions was a competitive sport, and the choices of what to publish were often made more on the basis of rarity than relevance. The list of what the Society proposed to print is a challenging read. There were "Popular treatises on the science of the Middle Ages," as well as "Treatises in Geometry written in England during the 13th and 14th centuries." A collection of letters by "eminent mathematicians, of the Seventeenth Century," before the publication of Newton's *Principia* is intriguing enough. Farther down the list of holdings, however, things get murkier. The Society also planned to print "an English tract on the making of oils and medicinal waters, from a MS. of the Fourteenth Century," and a collection of tracts on lithotrity, the surgical method of pulverizing gall stones.

In any event, the Society managed to produce only two slender volumes marked with its imprint. The first was the collection of correspondence, edited by Halliwell himself, which appeared as *A Collection of Letters Illustrative of the Progress of Science in England from the Reign of Queen Elizabeth to that of Charles the Second* (London, 1841). Halliwell's partner Thomas Wright edited the second, which consisted of popular treatises from the Middle Ages.

One contemporary reviewer remarked that while Halliwell's book contained little that was truly novel or groundbreaking, it was comprehensive and surprisingly entertaining.[15] De Morgan, reviewing the work anonymously in the *Athenaeum,* was more enthusiastic about what he saw as a foundation for future scholars to work upon. What may look like a mass of undigested material was the raw matter needed to fuel a new history of science. The "history of English science is not yet written," he lamented, even as "foreigners have their histories in which the rise of the subject is traced from the beginnings."[16]

An early promoter of the history of science, Augustus De Morgan believed that "the most worthless book of a bygone day is a record worthy of preservation." He thought that Newton's personality and beliefs were irrelevant to an appreciation of his scientific achievements. © National Portrait Gallery, London.

De Morgan was unusual in his zeal for the more obscure products of science and mathematics. He too understood the challenge that the old materials presented; it was a lot to ask of a single scholar to have the technical skills (such as the ability to decode ancient handwriting) and enthusiasm necessary to tackle the work. "When will it happen," he wondered, "that a paleographist is also a mathematician, with enough energy and leisure both to work the ore and the metal?"[17] De Morgan hoped that the Historical Society of Science might supply an institutional replacement for the heroic individual, as yet unforthcoming, who might carry out such a task.

But the fledgling organization was not up to the task. The Society expired from a lack of funds and a lack of leadership. Halliwell was as ambitious as ever and soon set his sights elsewhere. He made a visit to Sir Thomas Phillipps and almost immediately began wooing Phillipps's daughter. Despite the vehement protestations of Phillipps, who thought Halliwell an inappropriate husband, the two were married, eloping in 1842 and gaining the bibliophiliac son-in-law the hefty surname of Halliwell-Phillipps. Phillipps devoted much of his energy over the next thirty years to demeaning his son-in-law in print, accusing him of theft and misrepresentation, and not without cause. Halliwell-Phillipps was accused in 1845 of having stolen scientific papers from Trinity College seven years previously, when, as a student, he had had access to the locked shelves. He claimed innocence, and the scandal reached national heights. Prime Minister Robert Peel weighed in in his capacity as trustee of the British Museum, where the purloined manuscripts had ended up. It was never formally proven whether or not Halliwell-Phillipps had stolen the papers, and they remain at the British Library, not Trinity College, to this day. The irrepressible Halliwell-Phillipps went on to continue to make a good living buying and selling manuscripts and was active in donating materials to various libraries.

At the same time that Halliwell-Phillipps's archival transgressions were being investigated, Guglielmo Libri, who had been a corresponding member of the Historical Society of Science, was busy stealing a massive number of books and manuscripts from provincial libraries in France. He unsuccessfully tried to sell the collection to the British Museum, which declined after being tipped off that it was probably stolen goods. Surprisingly Libri was offered a job by sympathetic managers at the Museum when he arrived in England soon after, having been forced to flee France (with eighteen crates of his most valuable books and manuscripts) when his crimes were uncovered.

While the Historical Society of Science collapsed from lack of direction and funds, and the bibliomaniac crimes of men such as Halliwell-Phillipps and Libri cast a shadow on the organization, there remained other, more respectable avenues by which to get close to the origins of great thoughts. In 1838 Stephen Peter Rigaud, who, like Edmond Halley before him (and Baden Powell after him), was Savilian Professor of Geometry at Oxford, brought out a volume of high scholarly intent. This *Historical Essay on the First Publication of Sir Isaac Newton's Principia* did what Brewster and Biot had not managed to do. It took very seriously the evolution of Newton's thought as he prepared the first edition of his masterwork. It was based on much material that had never appeared in print before (some of which Rigaud included in a lengthy appendix), drawn from the collection of the Earl of Macclesfield (who had inherited Newton manuscripts in the possession of Newton's friend and frequent correspondent on mathematical matters, John Collins) and material held by the Royal Society.[18]

Rigaud's second book in the field, published posthumously in 1841, was a massive edition containing nearly a thousand pages of letters between men such as Isaac Barrow (Newton's predecessor as Lucasian professor), Flamsteed, and Newton, which Rigaud had transcribed himself from the Macclesfield collection. Much of it was famous but rarely seen material published in the "Commercium Epistolicum," a collection of letters pertaining to the dispute over calculus between Newton and Leibniz and published by the Royal Society in 1712. In his preface Rigaud's son reiterated that there was no better way to understand "the stages through which men's minds advance in the progress of discovery" than by examining their correspondence, a "written transcript of their occupations, their thoughts, and their difficulties."[19]

Rigaud's books display the values of professional history that remain familiar today: the need for completeness (all the letters were published so that "imperfect extracts" could be avoided) was to be balanced against the need for clarity (orthography was regularized and spelling modernized), since the point was to "throw light on the scientific progress of a remarkable period, rather than to produce facsimiles of the correspondence of individuals."[20] Rigaud exerted an influence but simultaneously strove to make his evidence available to his readers so that they could form their own conclusions. He understood the need for accuracy and a sense of proportion when making editorial decisions.

Baily, Halliwell, and Rigaud were not alone in attempting to move the history of science forward. Much that bore directly on the reputation of Newton continued to appear. In 1850 Joseph Edleston, a fellow of Trinity College, Cambridge, published an edition of documents and letters drawn from the college archive. Alongside correspondence between Newton and Roger Cotes concerning the preparation of the second edition of the *Principia* for the press, Edleston included the records of Newton's departures and arrivals at the college (recorded for all fellows at the time), a list of his lectures delivered as Lucasian professor, and Newton's dining expenses for a portion of his Trinity years. All of this detail about Newton's daily life, mundane as it seemed, could prove meaningful in the context of the new scientific history. Edleston remarked that many years' worth of missing dining hall ledgers would have been "indispensable for a correct history of the discovery of the new calculus and of the true theory of the world."[21] He used the new sources he did have to poke holes, as Brewster had attempted to do, in Biot's claim about Newton's mental crisis, noting that while Huygens implied that the breakdown had occurred in London, the Exit and Redit books at Trinity showed that Newton was never away from the college for more than a fortnight during 1692 and 1693.[22] The detailed marshaling of evidence continued.

Augustus De Morgan's own *Arithmetical Books from the Invention of Printing to the Present Time* (1847) was one of the earliest attempts to provide a complete catalogue of all such books. De Morgan addressed directly the vexed problem of obscurity and why readers should concern themselves with such esoteric topics as, for example, science in the medieval period. Wasn't science about the present and, more important, the future? What possible use was there for a history of science? De Morgan made explicit the link with astronomy, implicit in the literary avocations of astronomers like Baily and Rigaud: "The most worthless book of a bygone day is a record worthy of preservation. Like a telescopic star, its obscurity may render it unavailable for most purposes; but it serves, in hands which know how to use it, to determine the places of more important bodies."[23] In the hands of scholars, even most arcane points of reference could help to map the grander constellations of thought in the history of science. Nothing need, or indeed should, be discarded.

With that, De Morgan dismissed the criticism that could so easily be levied upon the seemingly impenetrable products of the printing societies. To that end, in 1846 De Morgan jumped in with what his wife, Sophia, proudly called the second work in English (after Baily's *Account*) to show Newton's "weak side." Baily's discovery of new correspondence had forced the issue, and "justice" for Leibniz, Flamsteed, and even Whiston demanded that the matter be even more thoroughly aired. In "Newton and Flamsteed," De Morgan brought to bear what was—again according to his wife—"a power of research possible only to one who was fully master of the history of Mathematical discovery."[24] In other words, De Morgan was the very man he himself had summoned in his 1841 article, a man who had the skills to mine both the "ore" and the "metal" of this historic-scientific puzzle. He advocated a kind of muscular history of science, unafraid to confront the weakness of its heroes and to recognize their strength in spite of the natural human failures. He noted that the "great fault, or rather misfortune, of Newton's life was one of temperament; a morbid fear of opposition from others ruled his whole life."[25]

De Morgan strongly rejected Newton's claim for the right to retreat from the litigious sphere of natural philosophy. "What is scientific discovery except filing a bill of discovery against nature, with liberty to any one to be made a party to the suit?"[26] De Morgan did not shy away from conflict, and he refused to exempt Newton from what he saw as the necessarily combative and communicative field of science. Discovery for De Morgan comprised both a natural philosophical and a social aspect: the scientist had to both find the thing out and to share it. Newton resisted the second component of the contract, and his contemporaries therefore stepped in to fill the unacceptable breach. "A discovery of Newton was of a twofold character—he made it, and then others had to find out that he had made it," was De Morgan's summary. The point was wittily made but serious. "To say that he had a right to do this is allowable; that is, in the same sense in which we and our readers have a right to refuse him any portion of that praise which his biographers claim for him."[27] The public's right to full discovery extended not only to natural knowledge but to knowledge about the lives of the knowers themselves.

In 1855 David Brewster finally produced his two-volume 1,032-page book, entitled *Memoirs of the Life of Isaac Newton*. Brewster's second Newton offering—finally available over twenty years after his first—changed the field of debate about Newton yet again, for he stoically included a raft of new documentary material, not all of which was easily squared with the model of Newton he wished to promote, but which, in an age of so-called scientific history, was deemed a necessary stiffener to a properly scholarly treatment of the great man. Contemporary reviewers immediately picked up on this. Baden Powell wrote that the delay between the publication of the 1831 volume and the 1855 work could at least partly be attributed to Brewster's having been "worthily employed in a laborious search into original documents not hitherto examined." It was worth waiting for a biography of Newton commensurate in size to his status, for, as Powell put it, "the history of Newton is in a great measure the history of science."[28]

Powell was right to draw attention to Brewster's use of new material. Brewster himself began his biography of Newton by reviewing the history of previous biographies, noting both the limitations of Bernard de Fontenelle's *Éloge* and the failure of John Conduitt to publish his own planned life (adding, somewhat ungraciously, that the "undigested mass of manuscript which he has left behind him" did not lead one to regret that he had given up on the idea).[29]

Predictably Brewster found the betrayal by both Baily and Flamsteed particularly galling. He bemoaned what he considered the

David Brewster in the 1860s. He fought long and hard to defend the reputation of his idol, Isaac Newton, against the madness and the meanness that Biot and Baily had revealed. © National Portrait Gallery, London.

lamentably wide circulation of Baily's *Account*. While he noted that only those 250 copies had been printed (at Admiralty expense, he railed), they had been distributed to precisely those influential readers who were best positioned to save or damage Newton's reputation. In response, Brewster announced that he had finally laid his hands on Flamsteed's letters to Newton (not included in Baily's account), letters that enabled him to "defend the illustrious subject of this work against a system of calumny and misrepresentation unexampled in the history of science."[30]

The matter of Newton's religion vexed Brewster, but he thought it pointless to hide from the public "that which they have long suspected, and must have sooner or later known," though he admitted to having "touched lightly, and unwillingly, on a subject so tender." Rather than trying to dispute or justify Newton's beliefs, Brewster decided to publish the relevant manuscripts and let his readers make their own judgment, urging them, however, to recall that "by the great Teacher alone can truth be taught."[31] Only God himself, "at his tribunal," argued Brewster, could explain how so much anti-Trinitarian chaff could have mixed with the golden wheat that was Newton's science. The legal metaphor remained apposite. In his review Powell referred to the "testimony" of one character who appeared in Brewster's "witness box."[32]

For Brewster, the question of Newton's faith was paramount. And he was not alone. Even in the middle of the nineteenth century, several troubling issues continued to surround the question of Newton's religion. Had Newton held a lifelong interest in theological matters or had it been merely the product of his dotage, the feeble imaginings of a weakened mind? Had he held beliefs that were irreconcilable with orthodox Christianity? Were his theological interests linked in any way to his breakdown of 1692?

Brewster was assiduous in bringing new evidence to bear on these questions, even if the answers sometimes disquieted him. He had seen with his own eyes the extent of Newton's work on alchemy, much of it notes on other writers or copies of their texts and indicated as much, though he could not understand "how a mind of such power, and so nobly occupied with the abstractions of geometry, and the study of the material world, could stoop to be even the copyist of the most contemptible alchemical poetry." Brewster put it down to the "taste" of the time in which Newton had lived, and left it, unsatisfactorily, at that.[33] More problematic still for Brewster were the new clues to Newton's theological beliefs that he dutifully published, including letters between Locke and Newton dating from 1690 on the subject of revelation and the apocalypse. "I should be glad to have your judgment upon some of my mystical fancies," wrote Newton to Locke. "The Son of Man (Dan. vii.) I take to be the same with the Word of God upon the White Horse in Heaven (Apoc. xii.), for both are to rule the nations with a rod of iron; but whence are you certain that the Ancient of Days is Christ? Does Christ sit anywhere upon the throne?"[34] Newton's impassioned questions did not meet with an immediate reply, and he was forced to write again, reminding Locke in a subsequent letter that he had not yet responded to his earlier question.

Brewster went on to discuss the contents of what he considered Newton's most significant theological writings, the *Observations* and the posthumously-published manuscript of Newton's "Historical Account of Two Notable Corruptions of Scripture," which contained the most problematic material. In that document Newton had left compelling evidence for his lack of belief in the Trinity, questioning the authenticity of the letter from John (in chapter 5, verses 7–8) in which it is written, "For there are three that bear record in heaven, the Father, the Son and the Holy Ghost, and these three are one." Newton's conclusion, faithfully reported by Brewster, was that this text was a "gross corruption" of the original scripture.[35]

Despite what looked like compelling evidence for Newton's heresy, Brewster was adamant that "this opinion is not warranted by any thing which he has published."[36] While he grudgingly admitted that, as De Morgan had shown, some of the phrases used by Newton in his writings were unlikely to have been made by a believer in the Trinity, he nevertheless valiantly insisted that the most this warranted was a mere suspicion of Newton's heterodoxy.[37] Brewster relied on some dubious reasoning to maintain his increasingly untenable position about Newton's beliefs. One was the distinction between what Newton had published and what he had kept private, most notably the writings Brewster had seen in the Portsmouth papers at Hurstbourne Park. Another was to insist on the uniqueness of Newton's belief. There are, Brewster argued, "various forms of Trinitarian truth, and various modes of expressing it, which have been received as orthodox in the purest societies of the Christian church."[38] It was possible, suggested the increasingly desperate Brewster, that Newton had simply developed his own form of truth, which, if closely examined, would be revealed to be as devout and acceptable as any true Christian belief.

While it pained Brewster to consider the possibility of Newton's anti-Trinitarianism, he believed it was his duty to investigate the matter as far as he could. The matter remained, even in the mid-nineteenth century, acutely relevant. While the 1813 Doctrine of the Trinity Act had formally extended toleration to those holding beliefs similar to Newton's—they could now legally call themselves Unitarians, serve as legal guardians, and hold civic office—prosecutions for blasphemy continued under common law procedures, and many were imprisoned for expressing their beliefs publicly. Ultimately the best evidence Brewster could find in defense of Newton was negative. As his daughter put it in her biography of Brewster, "He could not find any word or note in Newton's writings that could prove him to be beyond what is, I believe, called an 'advanced Arian.'"[39] Arianism, a form of anti-Trinitarianism, at least had roots in the early Church. This wasn't a great result, but it was the best that Brewster could do. According to his daughter, Brewster found evidence that Newton believed that Jesus was the son of God, but "there is never any recognition of His equality with the Father." Brewster never entirely lost hope, and, as his daughter put it, clung to the fact that there was "no distinct declaration of Newton's rejection of the doctrine of Trinity."[40] Nowhere did Newton ever state unequivocally that he did not believe in the Trinity, but neither did he make any robustly reassuring claims about the equivalence of God the father and Christ the son.

All of these rhetorical gymnastics were necessary because Brewster knew what Newton really believed. He had gone through the papers at Hurstbourne Park and had no question that they were the fruit of long and concerted labor—and that their author was not an orthodox Christian. In private Brewster was candid. He wrote Baron Brougham, the founder of the Society for the Diffusion of Useful Knowledge, that he had "ample proof" that Newton was an anti-Trinitarian. "No person who has seen his MSS. can entertain the slightest doubt upon this subject."[41]

In print, Brewster's final conclusion was that these writings offered an incomplete view of Newton's religious views. He praised Pellet, Conduitt, and Horsley for exercising a "wise discretion" in withholding these unfinished writings from the world and argued that had Newton had the time to complete them, they would have "exhibited more correctly and fully than the specimens we have given, his opinions on the great questions of Christian doctrine and ecclesiastical piety."[42] Ultimately Brewster fell back on the old reassurances of a typically British form of natural theology, praising the natural connection between Newton's investigations into the material universe—God's creation—and the divine will as manifest in scripture.

Brewster had served up the "facts" about Newton to an audience unconstrained by his pious framework and free to make their own interpretations. In general, reviewers focused on the personal details and the rehashing of old controversies. But both Biot and De Morgan criticized Brewster for not putting enough effort into unraveling Newton's scientific development. To them Brewster had squandered the golden opportunity afforded by his access to the Portsmouth papers to understand more about Newton's habits of mind. De Morgan and others immediately abandoned the image of the scientific saint to the rubbish heap of history, and in the process Newton was improved in their eyes. De Morgan had rendered himself ineligible for a Cambridge fellowship by refusing to sign the required statement of orthodox belief. Throughout his life he refused to assume the conventional emblems of membership. He declined an honorary law degree from the University of Edinburgh, never joined the Royal Society, and never visited either the House of Commons, Westminster Abbey, or the Tower of London. No follower of convention, De Morgan considered Newton's religious beliefs to be "as uncompromising and as honest" as his "philosophical ones."[43] De Morgan's matter-of-fact assertion was that the matter of Newton's heretical belief was "a vexed question no more." "We live," wrote De Morgan, "not merely in sceptical days... but in discriminating days, which insist on the distinction between intellect and morals."[44] It was no longer acceptable to expect a great scientist to also be a great man. While De Morgan applauded Brewster's change of heart from the writing of his first to his second Newton biography, he saw that it came at a great cost to Brewster, "who feels that he must abandon the demigod."[45]

In her memoir about her father, Brewster's daughter confirms De Morgan's suspicion that though Brewster felt he had to renounce the "demigod," he really yearned to defend Newton. It hurt Brewster to see his hero besieged. That more than a hundred years after his death Newton should still be attacked was a "personal grief and an English scandal." That was why Brewster had taken so long to write his second book on Newton, having devoted many of the intervening years to searching out "every proof and evidence" by which to defend Newton "from the charges against his sanity, his probity, and his justice, which were circulated when the hand and the tongue of the accused and his contemporaries were safely mouldering in the grave."[46]

De Morgan for his part refused to use the word *genius* to describe Newton, preferring the more mature-sounding *sagacity*. He took some pleasure in depriving his readers of any remaining hope of retaining the thrilling elements of the myth of Newton, chastising them for wanting "marvelous riddles solved, and some extraordinary feats of mind." "The contents of some well-locked chest are to be guessed at by pure strength of imagination," continued De Morgan, "and they are disappointed when they find that the wards of the lock were patiently tried, and the key fitted to them... by processes of art."[47] Science was not the province of divine inspiration or heroic moments of discovery but rather slow and patient work. De Morgan's "processes of art" are the practices of science—embodied, accumulated skills that are painstakingly acquired—rather than the pyrotechnics of cerebral force that so many wished to assign to Newton.

De Morgan similarly took pleasure in disburdening Newton of the mantle of religious purity. As a dissenter himself, he must have enjoyed the fun, and freedom, of showing what Newton really was: a heretic, just as De

Morgan and his fellow dissenters had been (and still were, in some quarters) considered. In a brief life of Newton published in *The Cabinet Portrait Galleries of British Worthies*, De Morgan wrote as openly about the matter of Newton's religious beliefs as anyone ever had. He presented evidence for both sides. On the one hand, Newton had friends who were unmistakably Arians (William Whiston and Hopton Haynes, an employee of Newton's, among them). On the other hand, he described Newton's own writings on the "two notable corruptions of scripture" and concluded that it was not possible to be certain exactly where Newton stood. While Newton had not given explicit evidence in favor of his anti-Trinitarianism in this piece, De Morgan pointed out that neither did Newton bend over backward to prove his orthodoxy. Given the severe penalties in Newton's day for being an anti-Trinitarian, it seemed fair to assume, suggested De Morgan, that someone who didn't take pains to disavow the heresy might actually have embraced it. Newton remained, despite all the buffeting, a "truly great man," according to De Morgan. The "blots of temper" need not obscure the deeper, more fundamental greatness. Moral virtue and intellectual achievement could—and indeed should—be separated. No "monster of perfection," as some biographers wanted of their subject, Newton remained "an object of unqualified wonder, and all but unqualified respect."[48]

Spurred on perhaps by his impatience with Brewster's limitations as a biographer, De Morgan wrote an additional semibiographical treatise on Newton. Rejected during De Morgan's lifetime by a publisher who thought it would not be of interest to the general public, it was published only posthumously, in 1885.[49] In an introduction written by De Morgan's wife, Sophia, we glimpse the kind of moral gymnastics De Morgan himself engaged in when it came to Newton.

In the work De Morgan considered the possibility that while living in his house Newton's niece had an illicit affair with the poet and statesman Charles Montagu, the First Earl of Halifax, whom Newton had taken under his wing at Trinity College. Voltaire had first suggested that Newton's appointment as warden of the Mint was a mere preferment as a result of his niece's relationship with Montagu. Brewster had denied both; De Morgan concluded that a secret marriage had taken place, so the affair was technically no longer illicit. Even the dissenting De Morgan had to admit it would be a moral step too far were Newton shown to have not only countenanced but knowingly profited from an secret affair carried out in his own house.[50]

On his deathbed in 1868 David Brewster, now duly knighted, still had Newton in mind. Speaking with a visitor about his faith, he remarked that he was still a devout believer. Remaining skeptical about of the word of God, he said, was "the pride of intellect—straining to be wise about what is written." He felt that, after a long and winding path, he had achieved the humility of his own ignorance. "How preposterous for worms to think of fathoming the counsels of the Almighty!" he exclaimed. And thus he was reminded once more of Newton and his famous dictum of humility: "I seem to have been only like a boy playing on the sea-shore, and diverting myself now and then with a smoother pebble or a prettier shell than ordinary, while the ocean of truth lay undiscovered before me!" But it was with sorrow rather than pride, finally, that he recalled Newton, for in the (very) final assessment, he could see only "how sad [it was] that Newton should have gone so far wrong" in his apparent embrace of anti-Trinitarianism. In the last, mystery was Brewster's great refuge and comfort. There were "secret things that belonged to God." That was plain enough to see, and to accept, even for a man who had spent his life trying, as Newton had already done, to see farther than others had seen before.[51]

Brewster was lucid to the end. Should anything be done with his scientific papers, he was asked. "No," came the emphatic reply. He had done everything a scientific man should do. Everything of worth had been published. There were, finally, no mysteries to Brewster's life. The dark and regrettable ellipses of Newton's life and the shadow manuscripts to his published works were to find no corollary among Brewster's works. The mysteries that Newton had left behind were allowed no place in Brewster's own life.

True to his word, Brewster—who had supported himself and his family throughout the course of a long and productive life by the abundant fruits of his pen—did not leave a substantial archive of personal papers. Just seventy letters are held in Edinburgh University Library, with a further fifteen at the University of St. Andrews.

# 7

# Wrangling with Newton

On July 29, 1872, John Couch Adams caught the 1:17 p.m. train from Cambridge to St. Pancras Station in London and transferred to Waterloo Station. There, amid the "great bustle" of the busy station, he met up with George Gabriel Stokes. The men were on a mission to bring Newton's papers back from Hurstbourne Park, the ancestral home of his descendants, to their original (and, many would argue, spiritual) home in Cambridge.

No one could have been better suited than Adams and Stokes to the task of exhuming the Newtonian relics from the darkness in which they had sat since Brewster's visit. Or so it must have seemed to the Cambridge officials who sent the famous scientists to courier the package of Newton's scientific papers, which the Earl of Portsmouth had generously offered to donate back to Cambridge, as he felt that they "would find a more appropriate home in the Library of Newton's own University than in that of a private individual."[1] The earl's sense of responsibility for the papers must have felt personal; his name was Isaac Newton Wallop.[2]

Adams and Stokes caught a train from Waterloo to cover the roughly seventy miles southwest from London to Whitchurch Station, in Hampshire, where they were met by Lord Portsmouth's carriage and driver. Lord Portsmouth himself arrived on the last train from London at nearly 8 o'clock, just in time for dinner. Lady Portsmouth retired early, but her husband, according to Adams, was very talkative that first evening. It was he who had written to Cambridge officials to offer up as a gift to the university a portion of the Newton papers, those touching on scientific matters, which had been passed down through the family for so many generations.

Isaac Newton Wallop, Fifth Earl of Portsmouth, generously donated the scientific portion of the manuscripts by his namesake and ancestor to Cambridge University in 1872. It would be sixteen years before they were catalogued. © National Portrait Gallery, London.

Stokes and Adams set to work first thing the next day and quickly realized what treasures there were amid the mess. On that first morning Adams recorded that he had come across some really interesting papers involving "atmospheric refractions" and "a few bearing on the motion of the Moon's apogee."[3] (In the evening, Adams immersed himself in *Goodbye, Sweetheart!,* a sentimental novel featuring a brash heroine and a moonlit boating expedition. He had surprisingly catholic taste in books, reading in botany, geology, history, and divinity, as well as fiction, which he relied upon especially "when engaged in severe mental work."[4])

Relations between the scientists and the earl were cordial. By August 2 the earl was more than ready to follow through on his generous offer. He reported to the vice chancellor of Cambridge—the administrative head of the university—that he had handed over "the

Newton manuscripts and two copies of the Principia 1st and 2d Editions corrected by Newton." Certain "fragments relating to mathematics" that Stokes and Adams had marked for careful examination would also be lent with the proviso that if they turned out to contain relevant calculations, those, too, would be donated to the university. The earl's generosity did not end there. He was also willing to lend (which he underlined) letters "from Eminent Men to Newton," which might be published if it were useful and if they threw "light on scientific questions."[5]

There were, however, limits to his liberality. The earl pronounced himself willing to hand over all materials related to science, but the papers relating to "personal matters" were to be loaned only. Those were "heirlooms" that he intended to keep. He expected the professors to be scrupulous in upholding these wishes. "I would rather cut my hand off," finished the earl, "than sever my connection with Newton which is the proudest Boast of my Family."[6]

The professors must have had their own hands full on the return journey, for, though Adams does not note it in his diary, they left Hurstbourne House not only with the scientific portion of the manuscripts but with practically all of the Newton material the Portsmouths owned. The Newton papers now embarked on their second great journey since Newton's death and a return of sorts—back to Cambridge, where Newton had packed up many of them when he had left for London in 1696 to take up his position at the Mint. For nearly 150 years, the papers had remained safely at Hurstbourne Park. They had been consulted remarkably few times: by Royal Society secretary Samuel Horsley in the late eighteenth century and by Baily and Brewster in the early nineteenth. Lady Portsmouth's explanation for why the professors had taken the whole lot was almost apologetic. She explained that the papers were in a "state of confusion," having never really been organized, and that in the bundles labeled "worthless" the professors had found "calculations of considerable interest & importance."[7] As Lady Portsmouth pointed out, the papers had never been fully sorted since Newton's death. A combination of their forbidding complexity and the lack of sustained interest had ensured that they remain shrouded, if not in complete mystery, then at least in a thick veil of modesty. Lord Portsmouth therefore thought it wise to allow everything to go to Cambridge. There was simply too much work to be done in sorting the papers. Even a hospitable host and hostess had their limits. It had become apparent during Stokes and Adams's flying visit that the operation of sorting the manuscripts would require months rather than days. No one suspected that the process might take years, or even decades. Stokes was an expert in viscous flow who had demonstrated how friction could cause droplets of water to be suspended indefinitely in the atmosphere. The Newton papers were to be similarly suspended—uncatalogued, partly on loan and partly in gift—for another sixteen years before the earl's generous gift, and its stipulated exception, would finally be honored.

Those sixteen years saw Cambridge remake itself as a location for an experimental and industrially relevant physics in which the role of Newtonian mathematical physics—still essential—underwent a complete revision. The pace and direction of change left Stokes and Adams straining to keep up. After long years of solitude in Hurstbourne Park, Newton's papers arrived precisely when their designated interpreters were struggling to come to grips with yet another brave new world: that of electromagnetic theory.

By the time Lord Portsmouth extended his invitation to the Cambridge professors, it was no longer deemed satisfactory that the literary and historical riches of the nation should rest—or, worse, molder—in the stately homes of the nobility. The justifications that an earlier generation had used for holding onto their literary hoards (the main argument being that the precious papers would be better conserved in the hands of the elite) were being challenged. Books and manuscripts should be moved from the aristocratic homes of the countryside to the centers of learning in Cambridge and Oxford and, increasingly, to London. New government offices there, such as the Public Records Office, sought to consolidate the records of the nation in one central location, where they would be accessible not only to the lawyers who relied on them to litigate property disputes but to growing numbers of historians seeking primary source material. Nothing less than a nationwide housecleaning was in order.

One indication of how far things had moved on was the establishment, in 1869, of the Royal Commission on Historical Manuscripts. While it had no legal authority to compel aristocrats and private institutions to open their doors, its mandate was clear. "There are belonging to many Institutions and private Families various Collections of Manuscripts and Papers of general public interest," the Commission asserted in its first report, which could prove "of great utility in the illustration of History, Constitutional Law, Science, and general Literature."[8] *Let us at them,* was the implication, *for they are, by rights, the property of the people.*

From the start, the Commission reported success of the barn- (or manor-) storming kind. In its first year it contacted 180 individuals and institutions and—in a telling indication of just how jurisprudential was to be its approach—appointed two barristers as its official inspectors of archives. In its second year the Commission honed its self-justification: "To those who are engaged in biographical, historical, or political researches no greater boon can be offered than well-authenticated information."[9] By the third report they were crowing that their success had surpassed their expectations. That success was properly a joint one in which both aristocrats and scholars could share. Everyone, it seemed, was behaving admirably. "The ready and liberal manner with which noblemen, gentlemen, and various public authorities, have thrown open their collections of manuscripts to the officers of the Commission" was equaled by the "eagerness" with which scholars and historians had welcomed the new materials.[10]

The discoveries were legion. A packet of letters found at Montacute House, the estate of the Phelips family, marked with the unpromising words "Law Papers" contained a collection of documents relating to the Gunpowder Plot.[11] Folios with marginal notes attributed (incorrectly, as it turned out) to Shakespeare were offered up for inspection by their owners.[12] Seventy-two original letters of Mary of Scotland, most in cipher, were discovered at Buckie, on the coast of the Moray Firth.[13] Original letters by Henry VIII, Philip, Queen Elizabeth, and Mary, Queen of Scots, were held by Lord Calthorpe at his estate in Norfolk.[14]

As the Commission's inspectors made their way around the private libraries of the land, lateral moves were also being made to get the archival wealth of the nation back into the national purse. They noted that one effect of their efforts was that collections had passed from private into public hands and "had thus become accessible to the historical student."[15] In the case of the Earl of Portsmouth, there was another reason for giving Newton's papers to Cambridge before the government inspectors had a chance to arrive. That reason was the Seventh Duke of Devonshire, William Cavendish. This was no trifling aristocrat. Cavendish had graduated from Trinity College, Cambridge—scoring second place on the extremely difficult final examinations in mathematics—in 1829 and went on to manage and increase the already extensive family holdings in both real estate and industry that were scattered widely throughout Britain. By the 1860s he was ready for loftier pursuits. In 1861 he assumed the role of chancellor of the University of Cambridge (a post he would hold for thirty years), and between 1872 and 1874 he served as chair of the Royal Commission on Scientific Instruction and the Advancement of Science (a committee on which Stokes also served). He would also provide Cambridge with a generous donation to establish the first university laboratory of experimental physics of its kind which was duly named after him. Cavendish had prompted Lord Portsmouth to open his doors to Adams and Stokes when they came calling that July in 1872, eager to secure Newton's manuscripts for the university before the government's Historical Commission could get to them.[16]

Paper was a valuable enough commodity in the nineteenth century that professors recycled their students' exam papers, turning them over and drafting their own scientific work on the back, thus preserving evidence, otherwise most likely to have been lost, of just how important writing was to the work of mathematical theorizing in Victorian Cambridge.[17] These habits of pen and paper have quite a bit to do with why the Newton papers went to Cambridge, why they languished for so long once they were there, and why, after sixteen years of cataloguing, they were deemed of little merit.

Those who scribbled hastily on those exam papers were students, above all, of Newton's mathematical physics. Though Newton had not cultivated a following during his own tenure at Cambridge, by the end of the eighteenth century the principles laid down in the *Principia*—and in particular the mathematical contents of that book—formed the basis for an intensely competitive system of testing at the university by which students were ranked in descending order based on their results on terminal examinations, known as the "Mathematical Tripos." (The origin of the term *Tripos* is uncertain, but it may refer to the three-legged stool on which students originally sat to take oral examinations.)

By 1820 a group of young students had managed to import the powerful innovations of continental mathematicians such as Leonhard Euler and Joseph-Louis Lagrange into the conservative university, reinvigorating what had become a moribund system crippled by the awkwardness of using Newton's mathematical notation for advanced calculations. But modernizing the notation did not mean abandoning Newtonianism. Instead these Cambridge modernizers understood this so-called "analytical revolution" to be based ultimately on the brilliant foundations set by Newton more than a hundred years earlier. It set the stage for mathematical physics to flourish at Cambridge, arguably for the first time since Newton himself had resided there. The updated

Tripos remained a paean to Newtonian subjects and Newtonian solutions in mathematics, astronomy, optics, mechanics, and hydrostatics.

There was always the potential for drift in such circumstances, and to address this William Whewell published a textbook edition of parts of Book 1 of the *Principia* in 1846. He noted that the great work had been studied at Cambridge ever since it was first published but that deviations from Newton's own text, increasingly common, were always of an "inferior texture" and "temporary currency." By hewing to the original text, students gained the "mental discipline, of having a fixed subject, of definite magnitude, which is to be mastered and understood."[18]

That fixed subject, and the material based on it, was hard enough that in the course of several grueling tests it could distinguish the great from the merely excellent with the brutal efficiency of a pin-sorting machine. Above the second wrangler was the senior wrangler, an honor enough to make its recipient an instant folk hero. His hometown threw celebrations, pupils at the local school enjoyed a holiday in his honor, and his achievement was reported in the national press. (When Joseph Larmor became senior wrangler in 1880 a "torchlight procession" paraded in celebration through the streets of Dublin.) Depending on the hopes they had harbored, lesser wranglers often carried the sting of disappointment or the glow of expectations exceeded far into adulthood. In 1874 a student who had been pegged to be senior wrangler came in at twelfth place instead, prompting a fellow undergraduate to comment that "it was as much as ruining his hopes for life: almost the sorest disappointment he could have."[19] Even men of such unquestionable caliber as James Clerk Maxwell, William Thomson (later Lord Kelvin), and J. J. Thomson carried with them through long careers the sting of having come in as second rather than senior wrangler.

The Tripos was designed to test bright young men to the very limits of intellectual endurance, and practically the only way to succeed on it was to train tirelessly. The test required a series of complex algebraic and geometrical manipulations that were possible to master only by means of pencil and paper. It was imperative to have a good coach, of which Cambridge had a legendary handful, most notably John Hymers, William Hopkins, and Edward Routh, whose students dominated the top three wrangler positions. Between 1865 and 1888 more than three-quarters of the top three wranglers were Routh's students, trained on his unique regimen of lectures (illustrated using a pedagogical innovation of the time, the blackboard), near-constant working out of practice problems, and repeated examination.[20] This one-on-one coaching, utterly unlike the sort of instruction offered in preceding centuries and certainly unlike anything Newton himself would have experienced, was by the 1840s absolutely necessary for success on the Tripos. These were coaches in every sense, for this culture of training was not narrowly mental. Physical fitness was a corollary to mental fitness, and the young men who prepared for the Tripos exercised daily. Swimming, rowing, walking, and running helped blow off the mental steam built up by intense study, and bouts of exercise punctuated both day and night. James Clerk Maxwell's roommate recounted with exasperation how "from 2 to 2:30 a.m. he took exercise by running along the upper corridor, *down* the stairs, along the lower corridor, then *up* the stairs, and so on."[21]

Far from being mere recapitulations of old knowledge, the Tripos were occasions when new research might actually be accomplished. Questions were deliberately posed to urge students toward novel research areas. Frequently students did just that, broaching new research in the high-pressure arena of the examination itself. Writing out answers enabled students to rehearse what they knew while grappling with what they didn't; written techniques offered pathways through the thickets. It's hard now to understand how novel this was at the time. We take writing for granted as a feature of all learning and examination. But until the early nineteenth century examinations had been oral, requiring students not to generate new research but to demonstrate their mastery of subjects in a catechistic form. The move to written examinations paved the way to uncharted territories where the lines between learning and research blurred. Stokes and Adams were exemplary of those who had achieved remarkable success as the outstanding "Newtonians" of this generation, beginning with their mastery of the Tripos. So it was that when Stokes and Adams found themselves in possession of Newton's papers, they held not simply the relics of a scientific saint; they possessed evidence of the writing habits of the nominal father of the pedagogical system of which they were both the prime products and, by their middle age, the preeminent producers.

In the 1830s David Brewster and Francis Baily had treated the papers as a set of legal evidence that could be used to defame or defend Newton's moral character. What events did the papers verify or disprove? What were the facts of the matter, when it came to Newton's religious beliefs or his dealings with John Flamsteed? For Stokes and

Adams, the sorts of questions that the Newton papers could answer were educational. At Cambridge, Newton was not only a patron saint but a pedagogical ancestor. He had provided the all-encompassing set of theories that in 1870 could still serve as a comprehensive framework for undergraduate education in the sciences. The chance to understand the inner workings of Newton's mind had the potential to deepen and extend a system built in his image. Like the student examination papers unwittingly preserved in the archives of their teachers, Newton's private papers were evidence of his mind in action.

Newton had been famously coy about his own methods, suggesting that he had kept his true means of discovering the *Principia* private and had only cast them publicly in the language of geometry. The question was therefore whether he adhered to the rigorous, manly, and above all morally upright techniques of thinking that Cambridge undergraduates were coached to acquire. To answer this Stokes and Adams were forced to consider whether Newton himself should—or could—be held accountable to the techniques that were mastered in his name. The Newton papers had the potential to probe more deeply the shadowy divide between patient work and divine inspiration, offering the promise of settling not simply what Newton had done but how he had done it. Brewster and Baily had raised the matter in their responses to Biot, but the question had a special urgency at Cambridge, where the moral value of study was paramount. In that respect the Newton papers mattered for every undergraduate preparing for the Tripos and for what the Tripos itself stood for. Would the man who served as a model for *what* should be learned also reveal himself, through his private papers, as a model for *how* to learn?

Stokes and Adams, for all they had in common, differed greatly in the matter of their relationship to writing. Stokes was a prolific, lifelong writer who knew himself best through writing and who exerted an enormous influence on the world of science through his writing. Adams was famous for not writing things down and, in one instance, paid severely for it.

The pair were exact contemporaries and shared much besides their birth year of 1819. Adams was Cornish, Stokes Irish. Both counted as "provincials" when they arrived at Cambridge in the late 1830s, at the tail end of the controversy over Newton sparked by Biot's publication, during the brief years when James Halliwell's History of Science Society was active, and while Whewell, as master of Trinity, declaimed the virtues of a Newtonian education in his famous Lancashire accent.

By the end of their undergraduate years, Stokes and Adams each offered incontrovertible proof of their intellectual superiority and, to defenders of the curriculum, proof that Newtonian subjects could generate thinkers of Newtonian caliber.

Stokes had showed early promise in mathematics, providing corrections and improvements to many of his school textbooks. Unlike Adams, who was the son of a poor tenant farmer, Stokes came from a relatively prominent Irish family, which had already sent many of its sons to Trinity College, Dublin. He studied at Dublin and Bristol College before matriculating at Pembroke College, Cambridge, in 1837. A man of devout and largely unwavering evangelical Anglican faith, he often took notes on sermons while he listened to them. Although he had not studied calculus until he arrived at Cambridge, Stokes did not waste time. With the help of William Hopkins, the best-known tutor of the day, he prepared for the Tripos. In 1841 he achieved the remarkable feat of finishing first as senior wrangler and becoming first Smith's prizeman, an additional award for mathematical prowess. His reward was a fellowship at Pembroke College. There Stokes spent the early part of his career, performing research on the motion of fluids that earned him membership into the Royal Society and cementing his reputation as a leading physicist of his day. During the 1840s he was busy with research on viscous flow and the transmission of wave motion through an elastic medium, as well as writing several papers that reformed the science of geodesy, the study of the earth's shape.[22]

In 1849, at just thirty-one years of age, Stokes was awarded that most Newtonian of mantles, the position as Lucasian Professor of Mathematics at Cambridge University, which Newton had made famous. Five years later he became secretary of the Royal Society, a position he would hold for thirty years. Stokes held the Lucasian professorship for an astounding fifty-four years, transforming it from a formal sinecure into a teaching platform from which he helped to shape a full generation's worth of the nation's leading scientists.

For his part, Adams was the sort of undergraduate whose abilities haunted his fellow students. One remembered meeting him on arriving in Cambridge in 1839 and recalled his despair at realizing that the first person he encountered was someone "infinitely beyond me."[23] What made Adams extraordinary was his ability to work things out in his head. In this he was the exception that proved the rule. While all around him students learned to make judicious use of the wastepaper

George Gabriel Stokes at around the time when he and Adams were charged with cataloguing the Newton papers. He held the Lucasian chair—Newton's position at Cambridge—for fifty-four years, teaching generations of undergraduates the precepts of Newton's optics, mechanics, and mathematics. © National Portrait Gallery, London.

basket as they practiced the techniques for solving problems such as calculating the forces at work on a spinning wheel and a suspended axle, Adams was legendary for his ability to solve such problems without writing them down on paper until they were finished. Though Adams was coached, he seemed nonetheless to have a method that was all his own.

Stories about the odd undergraduate abounded. Adams's landlady reported that she occasionally found him lying on the sofa without books or papers by him or simply standing at a desk and staring at the wall. The only way to attract his attention, she maintained, "was to go up to him and tap him on the shoulder; calling to him was of no use." To his fellow students, Adams seemed Newton-like in his focus. One said that it was when Adams was lying on a couch, seemingly at rest, "that his brain-work was at the greatest."[24]

John Couch Adams in graduation attire, soon after he finished as senior wrangler in the Mathematical Tripos in 1843. Reproduced by permission of the Master and Fellows of St. John's College, Cambridge.

It is hard to know whether such stories are apocryphal, but Adams's performance on the Tripos, at least, is a verifiable triumph. Incredibly he scored twice as many points as the student who was second wrangler and did so easily. "When everyone was working hard," a fellow student remembered, "Adams spent the first hour in looking over the questions, scarcely putting pen to paper the while. After that he wrote out rapidly the problems he had already solved 'in his head.' "[25] Like Stokes, in 1843 Adams graduated as both senior wrangler and first Smith's prizeman. He too was granted a fellowship at his college, and he soon began work on a project that had presented itself to him as an undergraduate, one of the remaining "inequalities" in the Newtonian system: the unexpected deviation of Uranus from the path predicted for it by Newtonian mechanics. (Newton himself had made no such prediction, because the planet was discovered by William Herschel only in 1781.)

While Adams was engaged on this project, his peculiar relationship to writing would reveal its serious drawbacks. Written records were essential to scientific discovery. Without them, Adams would find out, it was practically impossible to discover anything.

There were two possible explanations for Uranus's deviation. The first was that Newton had been wrong. The second was that another body, as yet unobserved, was affecting Uranus's orbit. Adams decided to find out which was true. The plan involved extensive calculations to determine the path of an undiscovered planet of the right size and location to produce the deviations. Adams, like Newton, took pleasure in calculation itself. In this and in his choice of project, he was self-conscious in his allegiance to the greatest Cantabridgian of all. He had, his biographer recounted, both "the greatest possible admiration for Newton" as well as a mind that "bore naturally a great resemblance to Newton's in many marked respects."[26] By September 1845, two years after he had decided to try to solve the problem of Uranus's unexpected deviation, Adams had pinpointed the place in the heavens where a previously unobserved planet might be—or, as he was convinced, had to be—located. The calculating was not easy, but in retrospect it would look simple.

Written communication was much, much more complex. Adams knew that it was only a matter of time before others responded to the puzzle posed by Uranus's path. His next course of action was typical of a man for whom writing was anathema. His idea of making his finding public was, to put it mildly, obtuse. Though the details are somewhat obscure, he most likely recorded the astounding news of a possible planet onto a small, undated piece of paper that contained no indication of how he had arrived at the discovery.[27] He sent this paper to James Challis, the director of the Cambridge University Observatory. But what had worked on the Tripos—neglecting to show his work on the page—was not sufficient for the matter of discovering a new planet.

Challis urged Adams to share his results with George Airy, the Astronomer Royal and the man who had set the problem to which Adams was supplying the solution. Adams did so, but in such a meek and inconsistent way that it was later questioned whether he had, in fact, discovered the planet by that point. Instead of writing a formal letter announcing his methods and results to the Astronomer Royal, Adams stopped by the Royal Observatory, unannounced, on his way home to Cornwall. This was a highly unusual, not to say irregular, course of action to take, even for a senior wrangler. Airy was a busy and high-ranking civil servant. Adams was, for all his promise, still merely a young Cambridge fellow. In the event, Airy was not at the Observatory. Adams stopped by again in October and, after finding

Airy unavailable, dropped a brief summary of his findings through the Astronomer Royal's letter box. This document, like the paper he conveyed to Challis, was notable mostly for what it lacked. Though it contained the key detail of the hypothetical planet's predicted mean longitude, Adams once again had neglected to indicate how he had come to his findings or even to date the correspondence (essential in proving priority). Without such an explanation, it was impossible to assess whether Adams merited the use of valuable telescope time to verify his discovery.

How Adams had managed the niceties of scientific communication would come to matter, because while he was busy dropping unsigned and undated notes into the mailbox of the Astronomer Royal and neglecting to reply to Airy's reasonable queries thereafter, a Frenchman named Urbain Leverrier solved the puzzle and managed to arrange for a telescope to be pointed at the right portion of the sky, thereby actually bringing the new planet into view for the first time.[28] Had Adams answered Airy's response to his note, expressing his interest in Adams's "paper of results" on a "planet with certain assumed elements" and asking some follow-up questions, Airy might have started a search for the planet immediately. Instead, having received no reply, the senior astronomer chose to treat the odd communication of the young mathematician as "doubtful" and leave it at that.[29] The result was that the Frenchman won the race to find the planet that came to be known as Neptune.[30]

After the discovery by Leverrier, Challis wrote to reassure Airy that impolite as Adams's failure to reply may have seemed, it was not intentional rudeness: "I have always found him more ready to communicate orally than by writing." Challis went on to say that he had begun his observations of the new planet based on nothing except the small, undated piece of paper. Challis knew that although it was most likely true, it was almost unbelievable that a young man would let the prize of a new planet slip out of his hands for such a simple and strange reason as an aversion to writing. "The public," he admitted, "would hardly take such a reason as that I have mentioned to be the true reason for his not answering your question, and I fear a hiatus must remain in the History."[31] Into that hiatus much national pride disappeared.

Leverrier was awarded the Copley Medal by the Royal Society in 1846 for his discovery. The British had to admit, though it was painful to do so, that the French had beaten them to it. Truly, though, the credit did not seem to matter to Adams. The controversy over who should be counted as the true or at least co-discoverer of Neptune dragged on without him, throughout his lifetime and well beyond. By all accounts, Adams took no part in it.[32]

The person whose honor concerned Adams more was Isaac Newton. His resolution to calculate the predicted orbit of the unknown planet was based, as Adams's biographer puts it, on his "complete faith in the Newtonian law and in the results of his own mathematics."[33] Adams's conviction that Newton's inverse square law of gravitational attraction must be correct had led him to embark upon the calculations in 1843. Some had suggested that perhaps Newton's great law of attraction, which stated that the attraction between any two objects decreases by the square of the distance between them, did not apply at such great distances from the sun. But Adams saw it the other way around: "The law of gravitation was too firmly established for this to be admitted till every other hypothesis had failed." In the end the "discrepancies" that had called the law into question were given the "the most striking confirmation."[34] That was why Adams had inferred that another planet must exist. His inability to question Newton's achievement had given him the faith necessary to persevere in his solitary calculations.

That Adams had in fact made the calculations first did not matter when it came to assigning credit for the discovery on the international stage. While Adams professed not to care, his fellow Englishmen certainly had reason to lament that the great Newtonian had not developed the kinds of writing habits that were essential to scientific discoveries—and scientific careers. If it hadn't been written down—and communicated—then in a certain sense it had not been discovered. More particularly in the case of Neptune, failing to alert astronomers with access to telescopes to the right place to look for a planet had ensured that they did not find it before the French did.

Adams did not suffer unduly for his failings, but only because he did not seek fame. He soon turned his prodigious calculating skill to determining the motion of the moon, a problem even Newton had complained "had made his head ach & kept him awake."[35] Adams recalculated the equations, going beyond not only Newton but Laplace to uncover novel lunar discrepancies (explained only in the twentieth century as the result of the slowing of the Earth due to tidal drag).[36] During this period Adams supported himself with a series of fellowships at his college, finally taking the Lowndean Chair of Astronomy and Geometry in 1859. He might have continued thus had not Stokes, together with Challis, intervened and convinced him to take over the post that would be left vacant on Challis's retirement. The new position as director of the Cambridge Observatory brought plenty of unexpected changes: a move to the outskirts of town, where the telescope was located, and an introduction to Eliza Bruce, a friend of Stokes's wife. At the age forty-three, and after nearly twenty years of intensely solitary work, Adams found himself married and in charge of a large institution, the Cambridge Observatory. Perhaps unsurprisingly he delegated the running of the Observatory to an able assistant and instead occupied himself with lecturing on the very Newtonian subject of lunar theory at the university.

Adams had been officially excused from overseeing the publications of the Observatory, and by 1872, when he and Stokes made the trip to Hurstbourne Park, nothing had been published for thirteen years. Airy (still Astronomer Royal all those years later) fretted that "Cambridge is beginning to lose place."[37] Still the pace did not quicken. The observations from 1861 to 1865 were published only in 1879. Some thirty years after Adams had left a scribbled note at Airy's house, the elder astronomer still had reason to despair of the younger.

Unlike Adams, for whom words were an impediment, Stokes seems to have known himself only through writing. During his fellowship years at Pembroke, Stokes struggled to balance his proclivity for isolated study with the pleasures of what he called "domestic affection." Having met Mary Susanna, a daughter of the famous Irish astronomer and head of Armagh Observatory, T. R. Robinson, at a meeting of the British Association for the Advancement of Science, he proposed marriage and was accepted. But finding himself engaged, Stokes was beset by an almost paralyzing anxiety about his capacity for marital happiness. Writing was the only way he knew to find his way out of this awful predicament.

Stokes's letters to Mary Susanna ache with his desire to find self-knowledge in exposition. His honesty about his fearfulness is striking. He was plagued by doubts, and his highly developed conscience forced him to make those doubts painfully clear to himself and to her. "The happiness of deep affection outweighs in my mind the happiness of the scientific leisure which I give up," he wrote her at one in the morning on a Sunday in June just three weeks before their scheduled wedding, "but the happiness of the scientific leisure may outweigh mere milk and water affection."[38] Would the marriage be a good one, he fretted, or just a mediocre "milk and water" one?

Stokes wrote one extremely long letter to Mary Susanna, explaining as much about himself as he could because it was better that she love him "as what I am than as something else." Somewhat apologetically, he sought a scientific analogy. His missive "was for a letter something like Hofmann's methylethylamylophenylammonium for a word: I guess you never got a 55-pager before."[39] Stokes simply could not stop writing as he struggled, with devastating earnestness, to find his way through the dilemma.

Stokes was only too aware of the pain of "being alone, breakfasting alone, dining alone, taking tea alone, week after week, so as to be deprived of the healthy invigorating effects of social intercourse and the mutual interchange of ideas."[40] He knew he would benefit from a break from himself and his mathematical fixations. Nonetheless he was grappling, as Newton never had, with the potential for attaining true scientific insight within the confines of a marriage. Much was to be gained, but much could be lost. That Stokes felt himself to be in a conundrum at all reflects contemporary assumptions about how scientific insights—discoveries—were attained, whether through slow, careful, and incremental accumulation of ideas and evidence, the kind of thinking possible within the confines of a marriage, or the solitude and intense concentration of bachelorhood.

In the end Stokes chose marriage (and Mary chose to marry him). They had five children together, though they experienced much tragedy, losing two children in infancy and a third in early adulthood. For a man who had feared his capacity to live socially, Stokes lived a relentlessly social life, in which he emerged as the central figure in a vast network of scientific correspondence.

Colleagues, contemporaries, and friends by 1872, when they set out for Hurstbourne Park, Stokes and Adams had both claimed the right to call themselves not just Newtonians but Newton's closest contemporary instantiations. Stokes was the first man since Newton to hold the Lucasian chair and the presidency of the Royal Society (a position he held from 1885 to 1890) and to serve, as Newton had, as a member of Parliament for Cambridge. Adams had proved by his faith in Newtonian laws and his prowess in their application that great discoveries could still be attributed to Newton.

Both men possessed the mathematical skills necessary to claim direct intellectual descent from Newton, and it was duly granted to them by their contemporaries. At his death, Adams was considered England's "greatest mathematical astronomer Newton alone excepted,"[41] wrote James Glaisher, his friend and obituarist. And Stokes was judged by J. J. Thomson to possess, like Newton, "that rare but effective combination of the highest mathematical powers and the greatest experimental skill."[42] The lectures that Stokes delivered as Lucasian professor were considered essential preparation for the Tripos. His influence on education at Cambridge was overwhelming. The roughly twenty students in his course included about 80 percent of the top ten wranglers for each year until the 1870s.

While no one would argue that Stokes and Adams were not great scientists and great Newtonians, when they arrived to inspect the Newton papers at Hurstbourne Park, being a Newtonian did not mean the same thing it had in their youth. Science at the university was changing, remaking itself to meet the needs of a powerful empire requiring scientific expertise in such things as laying undersea telegraph tables and designing faster steam ships and safer locomotives. In the same year, the university established the Cavendish Laboratory, generously funded by the same Lord Cavendish who urged the Portsmouths to donate their Newton papers to the university. At the laboratory students used precision instrumentation to run experiments on the new, industrially oriented subjects of thermodynamics, electricity, and magnetism. Instead of training students to compete on a paper-based test using mathematical techniques, the Cavendish lab created students who would be prepared to apply their knowledge in the workshops of Britain, refining techniques for generating and harnessing energy. Headed successively by James Clerk Maxwell, Lord Rayleigh, and J. J. Thomson, the Cavendish helped transform Cambridge from a place where mathematical physics was largely accomplished with pen and paper to one where a combination of theory and laboratory experiment drove research in a range of natural sciences.

Stokes himself considered taking up the first Professional Chair in Experimental Physics as head of the new Cavendish Laboratory, but he was wary of the brave new world the laboratory represented. He wrote to William Thomson about his hesitation over lecturing on new subjects. "I do not think you need be at all deterred by 'electricity and magnetism,'" responded Thomson. "The lecturing on heat and light, or even on 'light' without the name 'heat' would, it seems to me, be in all respects a proper fulfillment of the 40-lecture part of the duty." Stokes remained unconvinced. He felt he was too old and that it would be unfair to block the promotion of younger men.[43] To his credit, Stokes welcomed the future, if from a safe distance. He contributed to the wider reforms occurring in Cambridge physics, which included broadening the Tripos to include the new subjects worked on at the Cavendish, and, more generally, contributed to scientific education in the period. But he never fully embraced or mastered the new physics of his protégés.

What Stokes did instead was to act as Secretary for the most important scientific journal of the day, the *Philosophical Transactions of the Royal Society*. In fulfilling that duty for more than thirty years, Stokes wrote and received a mountain of correspondence in his study in the house on Trumpington Road, and later Lensfield Cottage. This meant communicating with authors, soliciting referees' reports, and editing papers for publication. It was not unusual for him to send three letters in one day to a single correspondent on some pressing matter, and then follow up with a telegram to clarify an important point.[44] It is little surprise that, as his daughter recounts in a vivid reminiscence of her father, his office was awash in paper, a sea of correspondence, notes, and manuscripts that Stokes was never entirely in control of. Stokes had, his daughter noted, "two really wicked characteristics": he would not allow anyone to help him with his work, and "he kept every single thing he received by post, even advertisements." The combination led to a maelstrom of paper. His study was the subject of no little domestic strife. It was, his daughter remembered, "enough to drive any housemaid 'wild.'" Stokes stuffed his office with as many tables as he could, "with narrow passages between, through which to squeeze if you could." The tables were piled with papers more than a foot high. It was little surprise that Stokes couldn't find anything. Before he went to London, there were desperate hunts through the mess. He would start by resisting interference, but when the searches "grew more desperate," as when it was time to catch the train to deliver an important lecture, he accepted help. When Stokes was out of the house (in London for Royal Society business, for example), his wife took the opportunity to "fall upon these rooms, when clothes-baskets full of unnecessary matter would be removed."[45]

Imagining the scene, it becomes less incredible that the cataloguing of the Newton papers took as long as it did, and more incredible that it was ever finished at all. Stokes was a notorious procrastinator, failing to publish much of his own research and becoming infamously late in fulfilling many of his obligations. In addition to the busyness of his schedule and the normal level of chaos in his study, he seems to have developed a real aversion to the task of cataloguing, a chore that came to seem more and more onerous as the years passed. In his memoir of Stokes, Joseph Larmor noted his "dislike of prolonged uncongenial tasks." The older man, Larmor admitted, "was occasionally very dilatory about business, and would take quite a dislike to something

John Couch Adams in 1888, the year the Cambridge Syndicate finally completed cataloguing the Newton Papers. Adams was famous for working out complex mathematical problems in his head. Courtesy of University of Cambridge, Institute of Astronomy.

he had to do." Having been given Newton's papers to catalogue, Stokes "kept these precious documents so long that there was some anxiety as to whether they had been overlooked, and after letters had been written to him on the subject in vain, other members of his family had to be approached." Had he looked at the papers a few at a time he might have gotten through them "by degrees," but Stokes was daunted "by a large piece of work and quite began to hate it, naturally more so the longer it was deferred."[46]

Stokes was not alone in his dilatoriness. While Stokes foundered under the shoals of his multiple obligations and his lack of organization, Adams suffered the perfectionist's disease of not wanting to let anything go. Even his longtime friend and defender James Glaisher had to admit that Adams postponed publication "with the intention of effecting improvements in the processes and mode of representing the subject, or of attaining to an even more accurate result." For Adams, it was always possible to extend a line of calculations a few digits farther, and he was well-known to his friend for his perpetual claim when asked about a piece of work, "I have still some finishing touches to put to it."[47] The result was almost perpetual delay.

Adams had been more than willing to take on the cataloguing of the mathematical portion of the Newton papers. It was, Glaisher wrote, a

> difficult and laborious task, extending over years, but one which intensely interested him, and upon which he spared no pains. In several instances he succeeded in tracing the methods that Newton must have used in order to obtain the numerical results which occurred in the papers. The solution of the enigmas presented by these numbers written on stray papers, without any clue to the source from which they were derived, was the kind of work in which all Adams's skill, patience, and industry found full scope, and his enthusiasm for Newton was so great that he had no thought of time when so employed. His mind bore naturally a great resemblance to Newton's in many marked respects, and he was so penetrated with Newton's style of thought that he was peculiarly fitted to be his interpreter. Only a few intimate friends were aware of the immense amount of time he devoted to these manuscripts or the pleasure he derived from them.[48]

In their opposing ways, Adams and Stokes struggled with, avoided, and delved into the papers to which their specialties made them suited. What they found bore directly on their identity as Newtonians as well as on the foundation of the entire Tripos system.

8

# Newton Divided

Adams and Stokes were not the only men grappling with Newtonian homework. Stokes had received the optical portion and Adams the mathematical portion of the manuscripts, but the papers contained plenty of material that was nonscientific and which the earl wanted returned to him. Exactly what sort of nonscience these papers contained was not clear. Adams and Stokes were not the men to find out. Two other men were appointed to what was now called the Syndicate, charged with cataloguing and dividing the whole ungainly lot. Like Adams and Stokes, these two were Cambridge lifers. Henry Richards Luard and George Liveing spent virtually their entire adult lives within the university's protective (if at times impecunious) embrace, at Trinity and St. John's College, respectively. They were middling wranglers, having graduated fourteenth and eleventh. Neither man would pursue a research career of the sort that both Stokes and Adams were assured. Instead they found other roles for themselves.

As vicar of Great St. Mary's, Luard restored and renovated the church. As registrary of the university, he similarly set out to update its records. A "very prince of index-makers," he found a task to match his tastes in the gargantuan collection of documents relating to Cambridge's history. He rearranged, bound, and indexed all of the university documents in his care—a truly monumental undertaking. His assiduousness in carrying out his duties had an air of faithful penitence, for he displayed, it was said, an "almost painful attention to detail." From boyhood, Luard was a devoted book collector and was in a position in adulthood to indulge this passion in the form of scholarly study. He undertook as a hobby to edit a long list of works, including a translation of a medieval French life of Edward the Confessor and several entries to the *Dictionary of National Biography* (including the entry for bibliomaniac Thomas Frognall Dibdin). Trained as a mathematician but with a solid knowledge of the classics, Luard was a natural choice to help catalogue the portion of Newton's papers relating to history and religion, though his position in the Church naturally conditioned his response to them.[1]Liveing was the youngest man in the Syndicate and the most forward-thinking. Having finished a respectable if hardly spectacular eleventh in the Mathematics Tripos in 1850, he went on to take the Natural Sciences Tripos in 1851, the first year in which it was offered. There he placed topmost of the six students who took the exam, earning special distinction in chemistry and mineralogy, and embarked on a lifelong career in the former. In accordance with his more modest means, Liveing sought to accomplish for chemistry what Lord Cavendish would do for physics: to create a modern laboratory for the study and teaching of chemistry. He rigged up and paid for with his own money a chemistry laboratory in a centrally located Cambridge cottage. There he offered the first ever course of practical chemistry in Cambridge to medical students. Soon St. John's College recognized the value of his endeavor and built him a laboratory within the college walls, providing him a salary to run it. Liveing was a progressive in other areas as well, pushing the university into the new age of practical instruction and similarly urging Darwin to publish *On the Origin of Species.*

On paper Adams, Stokes, Luard, and Liveing made a comprehensive dream team for cataloguing the Newton manuscripts. For the first time the substantive divisions within Newton's papers—mathematics, alchemy, physics, and history—would be reflected in the expertise of the men examining them. Stokes's and Adams's qualifications were impeccable, and Luard and Liveing possessed substantial knowledge of their assigned subjects. This was progress, of a sort, in bringing Newton in all his facets back to life. In reality, procrastination and personal differences nearly crippled the project forever. More substantively the very makeup of the Syndicate, necessary not only to catalogue the papers but to divide them so that the nonscientific papers could be sent back to Hurstbourne Park, reflected the commitment to a divided—and hidden—Newton.

The men who were to judge the papers had already been judged, via the Tripos, according to a value system that deemed mathematical physics the highest form of knowledge. Stokes and Adams had come first, and Luard and Liveing, tasked with cataloguing the nonscientific papers, had fallen below the top ten. The Tripos ranking system, erected to sort students in the image of Newton the mathematical physicist, had ranked the cataloguers of his own papers, and by extension the contents of the papers themselves, even before they were sorted. That Newton himself might have thought differently about the merits of scientific versus nonscientific reasoning was unwelcome, almost inadmissible information. Thus the project to catalogue the papers was seriously handicapped from the start.

Though Stokes and Adams were nominally unconcerned with what was found in the nonscientific section, it was impossible to ignore that those papers might reflect badly on what sort of man Newton had been, just as the documents uncovered by Biot and Baily had. While the cataloguing project dragged on, the Historical Commission continued to gather papers from stately homes throughout Britain, including those of Hurstbourne Park, which it reached in 1878. What the inspectors found there revealed yet another side to Newton.

The inspectors sounded a bittersweet tone in their report, noting that "the number and value of the Hurstbourne MSS have been greatly lessened by the Earl of Portsmouth's recent gift of the scientific papers" to Cambridge. Still, a "remarkable collection of documents" remained in Hampshire. These manuscripts revealed, according to the barrister-turned-inspector J. C. Jeaffreson, not the Newton of genius nor the Newton of madness but Newton the petty bureaucrat: a man, the cataloguer of these papers was forced to conclude, who demonstrated "in a curious manner his inaptitude for clerical labour." Far from revealing Newton's position at the Mint to have been a mere sinecure, these papers, never before seen, showed that he had been "a painful toiler at petty and uncongenial tasks." The sheer number of rough drafts provided "impressive evidence" that had Newton been a better scribe he would have expended less time and effort on them. "He sometimes worked on as many as half a dozen several rough drafts before he was satisfied with the wording of an expression," was the cutting indictment.[2] Surprises about the nature of the man emerged from every corner.

Even Cambridge had its limits when it came to tolerating delay, and by 1887, fifteen years after it had been convened, the Newton Syndicate was coming under heavy pressure to finish the job of going through the Portsmouth papers. On March 8 of that year, Charles Taylor, vice-chancellor of the university and himself a mathematician and theologian, wrote to Adams to complain that Luard was not pulling his weight.[3] Three days later Taylor managed to extract from Liveing his contribution and reported that the order and division of the papers in the catalogue had been agreed:

Mathematics
Alchemy (or chemistry)
Chronology
Papers on Historical Subjects
Theology
Letters
Books
Papers returned[4]

It was also agreed, reported Taylor, that copies of the papers that would be returned should be distributed according to their subjects.

Five days later, on March 16, 1887, Taylor urged Adams to give as clear a report as he could. It was critical to have some sense of what the mathematical papers revealed about Newton's working methods on the *Principia*. "Put as much into the introduction to the Newton catalogue as you have time for," Taylor encouraged, "so as to bring out all the fresh points raised or settled by the MSS. For example, if they make it clear, or clearer, that he worked with fluxions and afterwards translated it into the old geometry."[5]

By July Taylor was losing patience with the bureaucracy, and in particular with Luard. In a letter to Adams he gave himself the opportunity to vent about Luard, who wouldn't agree to at least one meeting during the summer but wanted to wait until the October Term. "If we defer it till the October Term, it will not be finished up then and perhaps never will be! I suspect that you merely acquiesced in Dr Luard's proposal. But I hope that you will support me in the attempt to get the thing finished. The fifteen to sixteen years seem a fair allowance of time. I take no responsibility for further vacillation and delay."[6]

The Syndicate reported somewhat defensively that it had been a "lengthy and laborious business" to sort out the papers, confused as the documents were.[7] Among other things they had found evidence of Newton's extraordinary burst of creativity during the plague years of 1665–66, when he retreated to the relative safety of Grantham and began the process of discovery that led to the binomial theorem, the calculus, and a theory of orbital motion. Newton himself characterized this time of enormous creativity, when he "minded Mathematicks and Philosophy more than at any time since," as solitary and intense.

Whatever regimen he had been exposed to in Cambridge, he did not credit it with enabling the discoveries, which he had begun to make not in his rooms at Trinity but in his childhood home, deep in the Lincolnshire countryside. So far, the papers seemed to prove that Newton's saint-like devotion and visionary abilities did not constitute anything approaching a pedagogical method that others might follow.

The papers did, however, offer a few tantalizing and illuminating tidbits on the matter of method.[8] Among those on the lunar theory, though they were damaged by fire and damp, enough was preserved to give a general sense of his thinking. The manuscripts revealed moments when he changed his method of calculating because it led to what he called "practical inconvenience."[9] There was evidence in his calculations for determining how the relative motions of the Sun and the Earth perturbed the moon's orbit that he had changed his mind after reaching an unsatisfactory determination of the equation that described the lunar apogee, the point at which the moon was farthest from the Earth.

Here was confirmation that Newton had indeed worked by a process of refinement that inevitably included false starts and error. In this sense, Newton revealed himself to be less an otherworldly genius and more a figure with whom the Cambridge wranglers could identify, a tireless worker in the mathematical trenches, where progress was made by increments rather than leaps. Adams knew the feeling well. In 1853 he had published an important paper pointing out errors made by Laplace in determining lunar motion and promising to provide the correct calculations soon; it had taken him six long years to get the final numbers.[10] Here, in the papers, was evidence that Newton had worked just as hard to come up with his results.

The same sort of method was also in evidence in a series of calculations relating to the way a ray of light bends as it passes through the Earth's atmosphere. The Syndicate referred to a table of refractions Newton had compiled and that Halley had published in the *Philosophical Transactions of the Royal Society* in 1721, "without giving any idea of the method of its formation." His papers suggested that in coming up with the table Newton had not been guided by theory alone—proof of pure genius—but had worked out the path of a ray through an iterative process, what they called "an indirect method, making repeated approximations to the form of the path." Similarly the papers solved another mystery surrounding how Newton had determined what was called the "form of the solid of least resistance," a problem whose solution he had offered in the *Principia* without giving any clue to how he obtained it. Among the papers was found a draft of a letter written by Newton to David Gregory, Savilian Professor of Astronomy at Oxford, in which he clearly explains his method, a method that was, the Syndicate hastened to add, both "simple and ingenious."[11]

In their preface the Syndicate attempted to explain why it had taken so long for them to "examine, classify, and divide" the papers entrusted to them by the Earl of Portsmouth. It had proved a laborious business, as many of the papers were found in a state of "great confusion"—mathematical notes were found among theological treatises, and some of even the numbered leaves were out of sequence.[12] Though the editors stressed that Newton's mathematical studies practically ceased when he left Cambridge in 1696 and noted that there was little new to be discovered in the Portsmouth papers on that topic, they found new material in three areas of physics: lunar theory (no surprise, since this was Adams's specialty), atmospheric refraction (Stokes's), and the details of his determination of the form of a solid that experiences the least resistance when moving through a liquid. In these three areas, new things to be learned. Newton himself had indicated that lunar theory, as presented in the *Principia*, represented just a "specimen" of the subject.

If there was much of interest to be found in Newton's notes on lunar theory and the dispute over the calculus, the remaining papers on nonscientific subjects were, after sixteen years, given a summary dismissal. "Newton's manuscripts on Alchemy are of very little interest in themselves," the Syndicate reported, being little more than transcripts of other authors' work. When it came to alchemy, failure haunted the manuscripts. Far from being either simple or ingenious, Newton's writings on alchemy were complicated and ultimately—it was hard to deny—unsuccessful. The bulk of the material consists of notes on other authors and was not indicative, so said the Syndicate, of originality of mind. Worse, there was evidence of Newton's inability to accomplish what it seemed he had set out to do: trace "a connected system" of chemical relationships from the morass of material he had gathered from a group of clandestine alchemists, most of whom were known only by pseudonyms. From this secretive group Newton had received and made copies of documents containing coded alchemical secrets about how to harness and transform the vital forces of the earth's elements. These manuscripts remained in Newton's possession his entire life and were among the papers the Syndicate examined. They had formed the basis of Newton's unsuccessful project to derive fundamental principles that could render coherent the mess of highly symbolic and inconsistent material. They also proved that he had been a committed alchemist. On the positive side, Newton had evidently spent a great deal of time on what the Syndicate called "chemical" (as opposed to "alchemical") experiments, some of which, the examiners noted encouragingly, were quantitative.

Similarly the Syndicate declared that the historical and theological papers could not be considered of "great value." The sheer mass of papers indicated that Newton had spent "a great portion" of his later years "writing and re-writing his ideas on certain points of Theology and Chronology." The Syndicate was disdainful in noting Newton's propensity for writing out multiple drafts, indication of a wavering indecisiveness that contrasted unfavorably with the direct and vigorous style of thinking that Whewell had recommended to undergraduates. They dismissed this wealth of expression with the unconvincing argument that Newton had been captivated by the physical act of writing itself. It was too much to consider that he might have intended these works on the apocalypse and the early history of the Church to be read by others. "His power of writing a beautiful hand was evidently a snare to him."[13] The six or even seven drafts of the same essay were otherwise perplexing. Why, the reviewers wondered, was Newton so obsessed with getting these things right?

The members of the Syndicate all understood that writing was of critical importance, both for the solitary practical task of mathematical computation as well as for the social construction of knowledge. Writing accomplished this in two ways: by proving priority of discovery and by recording, in the form of correspondence, the debate and eventual consensus that allowed such discoveries to be collectively accepted. Writing provided the almost literal structure upon which the value of science lay, uniting the practice of thought itself (what is sometimes called "theory") on the one hand with the practice of social trust (what today we call "peer review") on the other.

The Syndicate had uncovered in the papers some evidence of Newton's method, or thought process, but by the second of these measures his form of writing was useless. Newton only reluctantly shared his work and did not tolerate peer review that he could not control. The man who supposedly had founded modern science had actively sought to conceal the pathways of his mathematical thinking and indeed developed highly critical, not to say cruel, ways of dealing with those who did not hew to his views. Among other things, he had refused to debate his theory of colors when challenged on his experimental procedures, waged a war against Leibniz for the right to claim priority in the calculus, and forced the publication of Flamsteed's star catalogue against his wishes. Newton had shared with no one the overwhelming majority of his writings on nonscientific subjects.

From the perspective of Adams and Stokes, it was a shame that Newton's papers failed to reveal more about his method. Worse was what they contained about his other interests. Repetitive and insular (part of the reason his chronology is so difficult to date is that it mentions almost no contemporary events), his papers were so constituted as to repulse efforts by men such as Stokes, Adams, Luard, and Liveing to reduce them to usable nuggets.

In sharp contrast to the earlier dismissals of would-be archivists such as Pellet, for whom the papers were overwhelmingly "foul" in the sense of being unfinished and rough, the Cambridge Syndicate found the opposite. In the care with which Newton had prepared his theological work was damning evidence of banality. Here was, implicitly, a statement about the kind of thinking the manuscript record should reflect. During the 1830s, 1840s, and 1850s Newton's then-closest intellectual descendants had been expressly trained to think with their pen in hand. Tripos examination papers recorded every step of their thought process (with the exception of Adams, of course). Newton's theological papers were damnably profuse but also, unlike those exam papers, suspiciously clean. They did not represent thinking along the lines that Stokes, in particular, would have recognized (i.e., in the Tripos papers he set and marked), nor did they reflect the necessary sociability of scientific correspondence. They betrayed at once an unseemly obsession and a sterile neatness.

Failing absolutely to note the salient, and unavoidable, feature of Newton's writing on history and theology—his passionate anti-Trinitarianism—the examiners revealed their own prejudices. As a progressive reformer, Liveing was the most likely among them to have given voice to the more radical nature of Newton's writings, but it was Luard who had been assigned to work through the history and the theology. Luard's Anglicanism probably meant that the best he could say about Newton's religious beliefs was nothing at all. Though the Syndicate had pointedly included Luard the humanist and Living the chemist, it was Stokes and Adams, the senior wranglers, who determined the Newton who was revealed by their catalogue. That Newton was the mathematical physicist and astronomer, the safe and reliable Newton of old.

If sterility and repetition in writing were suspicious in the realm of theology, there were also limits to the acceptability of mess in the mathematical calculations that represented the most valuable form of writing that Cambridge wranglers could produce. In 1883 George Darwin, son of Charles and second wrangler in 1868, devoted some of his inaugural lecture on being appointed Plumian Professor of Astronomy and Experimental Philosophy at Cambridge to an invective against the "slovenliness of style" that he witnessed infecting the Mathematical Tripos. Far from being a "mere question of untidiness," argued Darwin, the style in which examination questions were answered bore directly on their quality. "Minute attention to form" was essential. While admitting that great mathematicians had worked in different styles, Darwin asserted that "those who worked untidily gave themselves much unnecessary trouble." It was true, he noted wryly, that William Thomson wrote in a copy book brought out "at Railway Stations and other conveniently quiet places for studious pursuits," and James Clerk Maxwell used the "backs of envelopes and loose sheets of paper crumpled up in his pocket." For exceptional men such as Thomson and Maxwell, disorderly writing could be excused. For others, neatness of style accorded with clarity of thought. As ever, Adams was "a model of neatness in mathematical writing."[14] The student was advised to walk the line between the sterility of mere copying and the chaos of unruly notes.

In death, as in life, Adams and Stokes persisted in their different courses. Adams's papers were eerily clean. It was as if they had sprung entirely whole and finished from his pen. Having come of age in a Cambridge generation that had been trained to think out every step of a calculation with pen on paper, he had achieved the remarkable feat of writing only what he had arrived at satisfactorily in the safe confines of his head. Whatever mental mess there might have been (and evidence for it is scant) remained off limits for posterity. He left nothing to dent the reputation for pure thought that had haloed him since his undergraduate days. The result was an archive that felt almost artificial. So neat were the manuscripts that James Glaisher, who went through them after his death, found it "difficult to believe that they are not finished work that has been copied out fairly."[15]

Adams's published scientific papers and his unpublished lectures on the lunar theory were edited and reprinted by the University Press soon after his death in 1892. Adams was a passionate reader and created an impressive collection of early printed scientific books, most of which he donated to the Cambridge University Library (including a first edition of the *Principia*, one of four now held there). His astronomical books went to the Cambridge Observatory, and what remained went to Pembroke and St. John's. Two hundred books in the St. John's Old Library bear his name plate, many of them classics from the eighteenth and nineteenth century. His extant manuscript papers are also held in the college library at St. John's, including more than two dozen pocket diaries dating from between 1841 and 1890. Adams's hand is neat, almost tight, as he records on a daily basis the progress of his life, which was mostly the progress of his calculations.[16] Adams was buried in a cemetery close to the Observatory, and space was found for a medallion bearing his name and likeness near Newton's own resting place in Westminster Abbey.

On Stokes's death in 1903, the significance of his papers was immediately recognized by friends and scientists such as William Thomson (who had by then been knighted as Lord Kelvin) and Lord Rayleigh. Nonetheless the situation was mixed, for the bounty of his bequest was ballasted by its sheer overwhelming disorder. As one might guess from the state of his study, his papers were in considerable disarray. Making sense of them would demand "an organised plan of attack."[17]

Such a plan was quickly hatched by Stokes's many friends and admirers. No longer was it conceivable for the papers of a great scientist to lie fallow, and in Stokes's case there was energy and commitment to get the job done. Long before his death, he himself had started the process of editing and republishing his scientific papers, which appeared in three volumes (in 1880, 1883, and 1901). Given his notorious powers of procrastination, it is even more remarkable that these three volumes were finished before his death (though Rayleigh dryly noted that the third volume appeared a full eighteen years after the second, and a full fifty years "after the first appearance of any paper it contained").[18] Two more volumes appeared, under the editorship of Joseph Larmor (the man whose elevation to senior wrangler was celebrated by a torchlight procession), in the years immediately following Stokes's death. The battle to preserve and present his correspondence was commenced with similarly impressive speed.

Soon enough a striking discovery was made. What Stokes had left unpublished consisted mostly of jottings of arithmetical reductions and calculations.[19] There was almost nothing of true scientific interest; nearly everything of value he had already published in some other form or communicated to others by letter. The letters, however, more than ten thousand of them, contained an enormous quantity of scientifically valuable material. Unlike Newton, who had jealously guarded his ideas, Stokes had left much of his best work in the letters he wrote to friends and colleagues. His infamous procrastinating, which had robbed the scientific world of a promised treatise on optics and many other potential works, was to be finally, if only partially, redeemed by the publication of a set of correspondence in which the ideas he had shared with a few colleagues would become part of the public record. By 1902 all ten thousand letters had been sorted and arranged. They would appear in 1907, just four years after his death.

Newton's own scientific manuscripts were finally back in Cambridge, while the irksome material on chronology, history, and theology had been sent back to Hurstbourne Park. Meanwhile Newton still awaited a proper memorial in the form of a complete edition of his works. With the publication of the Syndicate's catalogue, that moment inched closer, but only just.

# 9

# English Books, American Buyers

After years of delay, by 1888 the Newton papers had finally been catalogued and separated. The scientific portion was finally available to scholars for study. Adams and Stokes had tersely noted some of their interesting features, but their catalogue had left much more to be discovered. And yet the papers, now deposited in the Cambridge University Library and freely available for the first time since Newton wrote them, would remain practically unexamined for the next sixty years.[1] The nonscientific papers, sent back to Hurstbourne Park, were also to enjoy another fifty years of quiet. The year of their sudden entrance onto the world stage, 1936, is the apex of this story, when the last of the Newton papers that had remained literally tied up at Hurstbourne Park for more than two centuries were finally set loose. Their entrance came in a moment of personal and financial turmoil for the Portsmouth family.

The decision to sell was made by Gerard Wallop (also known as Viscount Lymington), a direct descendant of Newton's half-niece Catherine Conduitt and her husband, John. As the heir to the family estate, Wallop was the inheritor of the Newton papers. He was also a passionate defender of the British soil and what he considered hallowed traditions of stewardship of that land. While not a member of the British Union of Fascists (founded in 1932 by Oswald Mosley), by the turbulent decade of the 1930s Wallop's politics were just as far to the right of the political spectrum. Wallop's mother was American and he had been born in the United States, but he saw no contradiction between his own mixed Anglo-American ancestry and the xenophobic, anti-immigrant, and anti-Semitic views he held.[2]

The library at Hurstbourne Park around the time the Newton papers were sold. Courtesy of the Hampshire Record Office.

Nor did Wallop perceive a tension between his love of adventure and travel and the exaltation of ancient traditions, rooted in the soil. While he argued on behalf of a vegetative stasis for the culture at large—peaceful pastures tilled since time immemorial by the same families—in his own life he roamed widely, following in his father's footsteps. Wallop's father, Oliver (later Eighth Earl of Portsmouth), had left England after an Oxford education for adventures in hunting and shooting on the American frontier. Though good breeding was as critical for the maintenance of cattle ranches out West as it was for the landed aristocracy of the Old World, Oliver found a way to plow a furrow all his own, drinking and shooting with Buffalo Bill as well as entertaining the kaiser's cousin. Gerard was raised on his father's ranch in Little Goose Valley, near Sheridan, Wyoming. Faced with his father's exuberant spirit but plunged, with the rest of his generation, into the horrors of trench warfare in the Great War, Gerard may have felt the need to cultivate an ideological stiffener for what might otherwise

Gerard Wallop, Ninth Earl of Portsmouth, in ceremonial robes and crown, around 1940. Faced with death duties and his own divorce, Wallop put the Newton papers up for sale in 1936. Courtesy of the estate of Paul Tanqueray/ NPG.

have seemed, in the postwar gloom, simply self-indulgent. Perhaps some of his intense feeling for the moral value of homeland derived from his having been abruptly expelled from his childhood Valhalla in America in order to attend boarding school in England. In any case, after a desultory tenure as a Conservative member of Parliament, Wallop resigned in search of more liberating pursuits. The measure of his political ambiguity can be taken by the fact that on his resignation, the Labour Party for his constituency asked him to run for them.

He declined the offer in favor of an experiment in what was carefully termed "leadership." He linked up with a man called William Sanderson, who had excavated certain "ancient" rites and rituals with which he proposed to reform the aristocracy, and its attendant servants, into fiefdoms of appropriately local, circumscribed extent. The problem written into this project, which was called English Mistery, was that while men like Sanderson and Wallop admired the racial and social views of Continental dictators like Mussolini, Horthy, and Hitler, they despaired of the European fetish for centralization. Mass rallies and hero worship were gaudy and destructive of precisely the local *terroir* that Wallop fancied himself as preserving. Written into the Mistery's program was the explicit assumption that hierarchy was a necessary corrective to democracy, with all that this implied for an underclass of workers and servants necessary to hold up the bulwarks of civilization. But these were always understood to be local hierarchies, local traditions. Power was absolute within the fiefdom; it didn't, by rights, extend beyond it. Above all else, for Wallop, it was the land and the traditional, conservative practices of husbandry that had maintained it that were to be admired and even worshipped.

Having cast off the seductive apparatus of totalitarianism and its power to co-opt not just the masses but the middlemen needed to run the estates, the Mistery left itself open to the rot of internal division and inertia. Wallop turned out to be cheating on his American wife of sixteen years. At the start of 1936 the scandal was made flagrantly public when notice of a *decree nisi* granted to the Viscountess Lymington appeared in the *Times* of London. The paper reported in pitiless detail that Lady Lymington had, in an "undefended" suit, requested the dissolution of her marriage on the grounds of Wallop's adultery with a Miss Bridget Crohan in July 1935 at a Chelsea address.[3]

Sanderson took the opportunity to attempt to gain back the upper hand he must have felt he had lost along the way. "You have before and after the commencement of the proceedings acted with such gratuitous indiscretion," he wrote Wallop, "as to embarrass the family life of the Stewards [the title of regional leaders of the Mistery] and others of your married friends and have thus displayed such irresponsibility as to deprive you of their confidence. Further they consider that the indiscretion has been so reckless as to make you a danger to the institution, for the protection of which you enjoy hereditary privileges."[4] But Wallop was canny, and shameless, enough to elide his personal scandal with the discord internal to the organization. When Mistery split in the fall of 1936, Wallop re-formed the group under the name Array and took most of its members with him.

It was in the context of these tumultuous personal circumstances that Wallop made the decision to sell the family heirlooms that Isaac Newton Wallop, the Fifth Earl, in particular had cherished, likening them to appendages of his own body. Following the announcement of his divorce proceedings in January, the next notice in the *Times* that bore Wallop's name was a Sotheby's advertisement in June, informing the public that the papers of Sir Isaac Newton were to be sold by order of the Viscount Lymington, "to whom they have descended from Catherine Conduitt, Viscountess Lymington, great-niece of Sir Isaac Newton."[5] "Isaac Newton as alchemist," ran an ad in the *Daily Telegraph*, "his theories on transmutation. Papers to be sold."[6] The combination of the financial pressure of his own divorce and the death duties payable on the £300,000 (£11 million today) estate of Beatrice Mary, Lady Portsmouth, Gerard's aunt and the widow of Newton Wallop, Sixth Earl of Portsmouth, had prompted Wallop to put the papers up for sale, though he could not hope to recover even a fraction of his costs by selling them. In April, Wallop sold Hurstbourne Park itself, including five thousand acres of "exceptionally good shooting and fishing" and a Jacobean-style house.[7] In addition to the large property at Hurstbourne Park, he sold cottages and smallholdings that were part of the estate. Compared to the psychic pain of giving up the homestead and the land, the pain of losing the Newton papers must have seemed relatively scant to Wallop.

It was a time when expectations and reality seemed to part company. In December of that year the king abdicated in order to marry twice-divorced American socialite Wallis Simpson. Despite what the staunchly royalist Wallop might have said about that publicly, he perhaps felt some private measure of fellowship with a man who could risk so much for love. He and Bridget Crohan were themselves married on August 14, 1936, the day after his divorce from his first wife was finalized.

And so it was that Gerard Wallop, the man who had vociferously proclaimed the values of Englishness above all else, would be the one to sell the Newton papers. Though it is striking today, that irony must have seemed a relatively small one at the time. English heritage of the sort that the Newton papers represented could not compete with the exigencies of Wallop's private life, nor with the larger economic forces that would bring so many country houses, libraries included, to the auction block during the Depression. Much larger quarry than the old papers was at stake—or, more aptly, under the gavel. A. N. L. Munby, working as a cataloguer for Sotheby's, serves as a guide into the vanishing world of the English country house library, circa 1930:

> The excitement of those visits linger still in my memory—the long journeys by train or car, the arrival at the lodge gates, the first sight of some beautiful house and the opening of the library door revealing

> hundreds of feet of mellow calf, russia and morocco bindings. There one worked for two or three days selecting and listing books worth sending to an auction sale in London, making unexpected discoveries at the back of the cupboards, seeking in vain for some missing volume of a rare work, but above all steeping oneself in the layers of culture represented by any library that has been maintained for several generations.[8]

In addition to long runs of *Punch*, the *Gentleman's Magazine*, and the *Quarterly Review*, often shelved in the billiard room where they might be browsed during a game, the typical country house library was filled with books germane to the maintenance of a large estate; gardening, agriculture, natural history, and sport predominated. Among them might lie unanticipated treasures. There was always the chance that something wonderful and forgotten would turn up. No matter how much the sale of the country houses and their contents might trouble those who regretted the loss of coherence and the "layers of culture" that Munby evoked, the trade in books did not consider itself crass or philistine. It was, as Munby patiently explained, "in the nature of things [that] the foundation of libraries of older books is dependent on the dispersal of other libraries."[9] Book collecting depended on a giant cycle of generation and decay, the turning wheel ruffling the pages on so many rare (and ever rarer) books whose like could not, after all, be conjured out of thin air.

Hurstbourne Park was not among the grandest of country homes, but the Portsmouth family shared the tastes of their contemporaries and peers. The Newton papers would have nestled in a library, or a strong room near a library, such as those that Munby describes visiting in the years of frequent auctions. The more general dispersal of country house libraries provides the immediate context in which the Portsmouth family, represented by Gerard Wallop, let go of the papers it had so earnestly conserved for more than two hundred years. Spurred by personal circumstances, this action was also part of a widespread transfer of property from the aristocracy to a much broader circuit of buyers, further extending the trade in books that had reached such dizzying heights during the bibliomania of the previous century. While that era's passion for trading books was full of speculative energy and the optimism that prices would rise, by 1936 fear rather than hope spurred the market. Families that had considered themselves untouchable by market forces found that they were not. The tokens of culture acquired over generations became subject to the ministrations of the market.

Trade had always been a feature—and indeed a lubricant—in the generation of great collections. Nominally genteel traditions of connoisseurship depended on it as surely as the coarser autograph manias of the nineteenth century. This was understood and accepted, even amid the jostling among the increasingly mixed class of book addicts (aristocrats, the nouveau riche, and the middlemen book dealers brought close by their passion). Value could accrue from such trading, even as it was unreliable. As long as the world of book buying and selling remained an essentially closed one, there was no danger of losing the books or of damaging the culture in which they held their meaning.

And so Munby could conclude, as he settled down to the rare delight of cataloguing the entire contents of a great private library, that he was simply contributing to the next round in a cycle that would repeat indefinitely. It was almost ecological, this churning up of objects from the sale lot of one steward to the bookshelf of another. His tone is neither elegiac nor mercenary. Rather it reflects crisp professionalism:

> I do not think that as I sifted the grain from the chaff in these libraries nearly forty years ago I had many feelings of regret at their dispersal. Sometimes indeed one was helping the owner to retain the best books. . . . More often, of course, he was seeking money, but whether it was to meet estate duty demands, re-roof the house, extirpate dry-rot, maintain his stud or grousemoor or pay his alimony, was none of my business. I had then, and as an academic I still have, a feeling that books are in large measure educational plant: and most of the libraries were certainly under-used . . . and I shared with many of my contemporaries the illusion that the supply of country house libraries was almost inexhaustible.[10]

There is nonetheless a certain ambiguity here. First, note that the books are (after all!) "educational plant," something to be *read.* Second, remember that Munby was there to organize a sale. Such sales did not, in fact, go on forever; the supply of country house libraries, as the supply of country houses, was not inexhaustible. Books were not primarily the stuff of scholarship (the plant of education) for Munby the cataloguer, however much he may have wished them to be so. They were a commodity.

Munby must have known that the titillating process of opening up private libraries to public view could not go on indefinitely. Nonetheless it did take the English aristocracy a long time to divest themselves—sometimes willingly, more often with intense displeasure—of their houses as well as the lives they had lived within. The destruction of the English country house, and of the aristocratic order on which it was based, was achingly slow. All told, it took about one hundred years for the process to run its course, from a rather dramatic beginning during the agricultural depression of the 1870s to a period of slower decline following World War II. As early as 1848 the contents of Stowe, the great ancestral home of the Duke of Buckingham and Chandos, were sold at auction in a desperate attempt to pay off heavy debts of more than £1 million (nearly £60 million today). The sale lasted more than forty days. Everything except the furniture was auctioned off—plate, books, manuscripts, prints, paintings, china, and sculpture. It was a blow to more than the duke and his family. The *Times* made its judgment clear: "When dynasties are falling around, and aristocracies have crumbled into dust, disgrace acquires the force of injury, and personal ruin is a public treason."[11]

The sale at Stowe was, everyone agreed, a result of monumental carelessness on the part of the duke. It was not, therefore, truly a bellwether of what was to come, and yet the sight of the mighty house laid open to the thousands who arrived on special trains arranged for the event to view its chattel prefigured a shifting in the social order. The realm of the possible had widened to include the destruction of what had seemed, if not eternal, at least so heavily bulwarked as to merit a stately decline rather than a mad rush. Instead the duke had brought himself down heavily, and the response was both to rubberneck and to regret, in the vague way that one regrets a natural disaster, that such a thing could happen.

By the 1870s there were larger forces at work than one man's heedlessness. An entire system of property, social and political power, taste and habit was eroding. It was a death by a thousand cuts, inexorable and lingering. Whether or not it was to be regretted depended largely on who stood to gain, ultimately, from the dispersal. And that was not as clear as it might seem. Changes in the fortunes of the landed gentry coincided with major shifts in the lives and working conditions of the masses. The Industrial Revolution and the agricultural depression that began in the 1870s robbed the houses and their owners of the income that had been necessary to support them even as it caused a huge exodus of workers and their families from the countryside to the cities. These workers found new rights alongside new indignities. The third Reform Act of 1884–85 extended the vote to roughly 60 percent of English men, diluting the political power of the great landowning families. The introduction of death duties in 1894, alongside other forms of taxation and the new surplus of grain emerging from the American prairies, further weakened the fiscal viability of the great houses. The trenches of World War I claimed many sons who otherwise might have managed and protected the great estates. The interwar period in which the sale of the Portsmouth papers took place represented the nadir. Land that had traded at £53 an acre in 1871 fell to between £23 and £28 an acre.[12] Debts acquired in the middle decades of the nineteenth century, when incomes were high, became overwhelming when rental income fell. The concatenation of factors meant that in the four years following the Great War ownership of land amounting to one quarter of England changed hands.[13]

It is hard not to slip into an elegiac mode when writing about this great divestment now, but the dismantling of the houses and estates was the dismantling of a system of class based on deep chasms of inequality. To regret the loss of the houses is also to regret the loss of a system in which the comfort of the few was purchased with the sweat of the many. The old sleight of hand by which such homes functioned as symbolic repositories of an English culture that belonged, nominally, to everyone was unconvincing in a world where so much that once seemed impossible had actually come to pass.

Land was a hard habit to break, and families looked seriously at other assets before putting the great country houses and their landholdings up for auction. Book collections were relatively easy to give up. Sons and grandsons rarely inherited their father's or grandfather's taste in collecting and, as Munby dryly noted, referring to celebrated English printers, "The loss of the Caxtons was, after all, much less conspicuous than the loss of the Van Dycks or the Gainsboroughs."[14] Books could help stave off the dissolution of an entire property. If they did not bring the same prices as the great Old Masters on the wall, they were also correspondingly less painful to part with. Spurred on by the Settled Land Acts of 1882 and 1884, which allowed heirs to sell possessions that had previously been protected by the wills of ancestors, the sales of country house libraries ramped up during the 1880s and 1890s. Manuscripts, often the records of a family's own illustrious history, were often saved until the very last, but when money became scarce enough, they too had to go.

This is not to say that in the brave new world there was no wealth, no wealthy. The great sales needed—and found—buyers, men who, if they did not precisely replicate the habits of the old families whose possessions they acquired, still saw value in their things. Some were American, some were Jews. Very wealthy men like

Ferdinand de Rothschild did not have the time or inclination to create a library through the slow, steady drip of connoisseurship. In February 1890 Sir Henry Ponsonby, private secretary to Queen Victoria, recorded acidly that during a dinner party at the home of James Knowles, editor of literary magazine *Nineteenth Century*, which was attended by William Gladstone, Lord Randolph Churchill, and Baron Rothschild, there was "much discourse on books...and how many would make a good library. It was generally agreed about twenty thousand of which Rothschild immediately made a note."[15]

The notes piled up as the newly rich studied their quarry and their quarry observed them. The sands were shifting. The objects were less valuable, in a way, than the guidance (implicit in that twenty thousand books number) on what counted as an entrée into polite society. Libraries were constructed with pipe organs built into the walls. Expensive silks covered whole flotillas of armchairs and divans, staged across a library's open space. Reading was not the only thing to be done with books.

In America there was new wealth and a new appetite for collecting the emblems of learning and civilization from the Old World. Dealers such as A. S. W. Rosenbach perfected the art of buying up the possessions of cash-strapped lords and selling them on to buyers who had the money and the desire to collect. The Philadelphia-based dealer purchased an unparalleled collection of early printed books from the philistine Second Duke of Westminster, whose passion ran to horse racing, not book collecting. The trouble with English dealers, noted Rosenbach, was that they failed to "buy great collections en bloc," nor had they "acquired the knack of securing great private libraries by private treaty." Rosenbach took advantage of this English distraction to acquire "wonderful things" from private sources.[16] He considered the £4,500 (£78,000 today) he paid in 1953 for just under four hundred books from the Duke of Westminster's library, most of them from before 1700, to be a bargain. The books were rare and getting rarer still, as American collectors clamored increasingly for European treasures.

Rosenbach sold to most of the biggest buyers in America, including the financier J. P. Morgan, "Copper King" William Andrews Clark, railway magnate Henry Huntington, and Standard Oil president Henry Folger. The books he sold to these titans of industry for the most part were literary, historical, and illustrated productions, the sorts of books that had traditionally been valued by English collectors too. Men whose fortunes were built with the tools of science and engineering purchased the letters of Walter Scott (Morgan), the works of the poet and playwright John Dryden (Clark), an elaborately decorated manuscript of Chaucer's *Canterbury Tales* (Huntington), and the first folios of Shakespeare (Folger). They bought with the same energy that went into their businesses, creating incomparable collections almost overnight and making Rosenbach rich. (His two boats were named the *First Folio I* and the *First Folio II.*) Twenty thousand books (the number Rothschild had noted as being the minimum necessary for a great library) was only the beginning for these men, whose libraries form the basis of what are today world-class research institutions open to scholars.

Rosenbach had one buyer with a taste for something markedly different and much more congruent with the scientific and technological basis for much of the wealth that enabled American collecting. This buyer made a fortune not in the coal mines or railways of a growing country but in the stock market, and he and his wife used that money to bring as many of the works of Isaac Newton to the shores of the New World—New England to be precise—as possible.

In Dogtown, Massachusetts, the boulders speak: *BE CLEAN. NEVER TRY NEVER WIN. KEEP OUT OF DEBT.* These hunks of granite are disconcertingly imperative when stumbled upon by the mountain biker or hiker exploring this overgrown park. Set on the outskirts of the old fishing town of Gloucester, Dogtown is a haunted sort of place, one whose loveliness is offset by its aura of careless decay. Prickly brambles run rampant, mosquitoes reign triumphant, and scattered building stones give evidence of hardscrabble habitation in years past by those who were not welcome in Gloucester, including the stray dogs who gave the place its name.

The boulders belong to a different kind of world, of well-swept floors and balanced account books, a world in which cellars do not degrade into rotten land and rose bushes do not grow unchecked. They seem as maddeningly certain of their message as of their permanence. And that is the point. The boulders speak the mind of Roger Babson, the millionaire investor and business guru of the 1930s. Babson hired stonemasons to carve twenty-three aphorisms into the hard granite of the large boulders scattered throughout the area, such as *BE ON TIME. WORK. STUDY.* "I am trying to write a simple book, with the words carved in stone instead of printed on paper," he explained in his autobiography, *Actions and Reactions*, the title of which is a direct reference to Newton's laws of motion and force.

Babson was not a man for wasted effort. He drew circles on his tablecloth to indicate where the plates should go. He dreamed of a mail-order clothing service, whereby shirts and shoes would be delivered at statistically predetermined intervals when their predecessors had worn out. He even had a plan for bed making that involved leaving the beds unmade during the day and arranging them at night, when a thermometer could provide exact guidance on the number of blankets required. Alas for Babson, his wife, Grace, refused these scientific housekeeping solutions, and the industrious Yankee had to content himself with applying his statistical methods to the stock market.

Today Babson is remembered as the prudent investor who founded Babson Institute, now Babson College, the first institution in America dedicated to educating undergraduates in business administration.

Convinced that Newton's laws of action and reaction underlay the financial markets, Grace Babson created the largest collection of Newton-related materials in North America. Reproduced courtesy of the Babson College Archives.

He also famously predicted the Great Crash of 1929 and made himself rich. Less well-known is the role he played in amassing the largest collection of Newton materials in America and of fostering research on gravity by leading theoretical physicists. Less well-known still is that Grace spearheaded the couple's investment in all things Newtonian. Roger was a man to declaim his thoughts, on boulders as well as financial charts, but Grace remains elusive. Her legacy today exists not in her own words but in the thousands of objects she collected that relate to Isaac Newton.

Finding nothing to contradict the notion that physical truth and economic truth share internal rules, Roger Babson applied a version of "to every action there is an equal and opposite reaction" to his Babsonchart. It worked like this: the Babsonchart was a composite chart, representing the price of various commodities and securities over time. To apply the law of action and reaction, it was necessary simply to calculate the area of the chart corresponding to a depression, or fall in prices, below some nominal "normal" line, and that of a period of prosperity. Once a depression area was equal in size to the preceding area of prosperity, another period of prosperity would be due.[17] In this application of statistics to investing, simple though it seems today, Babson was a pioneering spirit.

Babson had earned a degree at MIT and evidently set much store in engineering. But while he named his law after Newton and credited science with much invention, the ultimate rationale for the law of action and reaction in business was to be found in psychology and history, not physics. "In reality," explained Babson, "the business-cycle is the curve of man's attitude towards his neighbor and life."[18] In periods of prosperity, men tended toward "carelessness, inefficiency and unrighteousness." Business therefore "inevitably" deteriorated until, following a period of depression, men once more adopted the requisite habits of efficiency and hard work. Though Babson believed that at a fundamental level the two features of nature were linked—the tendency for countervailing forces to be ultimately balanced—his appropriation of the Newtonian mantle advertised a system that was not as economically naïve as it might seem. Historical insights offered a corrective to man's worst excesses. By analyzing the depressions of the past, it was possible to see the pattern and avoid the worst going forward. To avoid the contagious laziness of modern thought, in which no one thinks for oneself but merely follows the crowd, Babson advised "that business men look at things in a sensible light—in the

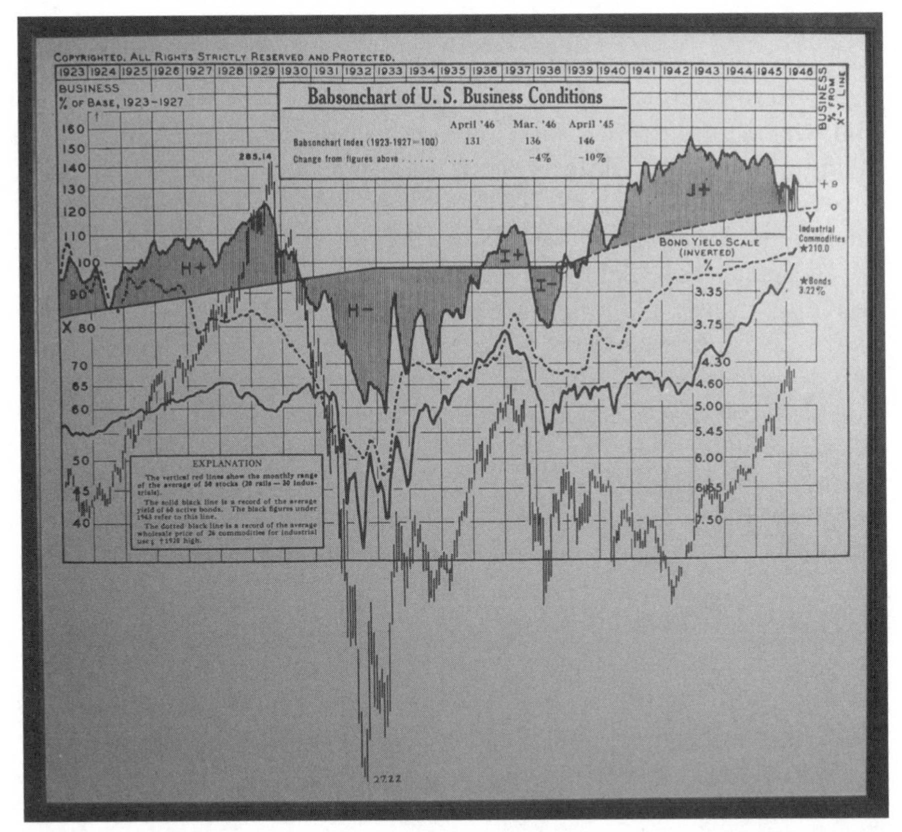


Roger Babson pioneered the use of statistics in investing and believed that Newton's laws of action and reaction explained the rise and fall of prices, as illustrated in his Babsonchart, with its areas of positive and negative growth. Reproduced courtesy of the Babson College Archives.

light of history."[19] What the light of history revealed in letters and other "human documents" from the past was the constancy of human emotion.

Roger Babson credited Newton with providing the basis for his successful business formula, but it was Grace Babson who really took to Newton. While the Syndicate's catalogue, and the manuscripts it described, sat unused in Cambridge, and while the country house libraries of England were being steadily dispersed, Grace Babson set about creating in America an unrivaled collection of Newton materials. Her interest in Newton stemmed from an introduction to his work as an undergraduate at Mount Holyoke College in the early 1890s. It further developed when Roger's advisor at MIT, Professor George

Swain, suggested that Newton's so-called law of action and reaction could be applied to nonscientific realms. Grace became convinced of the truth of Swain's suggestion, especially that economics was subject to the same fundamental rules as physics, chemistry, and astronomy. In 1906 the Babsons took the first of many trips to Europe. Their early trips were devoted both to selling the Babsonchart to English and continental bankers and to investigating the life and works of Newton.

In England the Babsons found a gratifying blend of tradition and ambition. "Their banking buildings were heated only by open fireplaces, which were also used for heating water for afternoon tea. There were few typewriters and almost no adding-machines. Quill pens were still being used, and speaking-tubes instead of interior telephones," Roger recalled in his memoir. Despite the sometimes archaic surroundings, the benefits of a long history were clear: "On the walls were the oil paintings of three or four generations who had always stood for integrity and courage."[20] As the home of Newton, England inspired Grace and Roger to start buying. They were well ahead of the curve in purchasing the treasures being catalogued and sold on from country estates. The Newton papers remained inaccessible, but there was plenty of printed material by or relating to Newton that piqued the interest of the Americans and could be had for surprisingly little.

In the years following their first research trip, Newton became Grace's abiding obsession. With the exception of what Roger called her "home duties" and looking after their only child, Edith, Grace devoted herself to Newton. According to Roger, she spent some time every day on her passion. The time was right for collecting: the library sales ensured a steady supply of new material; dealers were few, and even fewer were knowledgeable; buyers were also few, and fewer still could match both Grace Babson's focus, and, as the years wore on, her purse. Grace collected what is sometimes called Newtoniana, roughly meaning anything related to the man. She focused her attention on creating as extensive a collection as possible of editions of Newton's works, buying not only different editions of the same books (notably the *Principia* and the *Opticks*) but different copies of the same edition, which often had subtle but telling variations or annotations from important owners.

In 1926 Grace traveled, without Roger, to England on a buying trip for her ever-growing collection and stopped to lay a wreath at Westminster Abbey to commemorate the approaching 200th anniversary of Newton's death. The inscription on her wreath placed the Babsons in the long line of Newtonians but with a Babsonian twist: "His discovery two hundred and fifty years ago of the fundamental law of equal reaction is only today beginning to be applied in the fields of religion and economics."[21] On the same trip she bought from Foyle's, the booksellers on Charing Cross Road, a 1760 edition of the *Principia* for £30 (just under £1,000 today).

By 1935, a year before the Sotheby sale, Grace had acquired the largest collection of Newton-related objects and papers in America. This included Newton's own copy of the first edition of the *Principia* as well as Halley's copy (with notes by both men), a version of the second edition of the *Opticks* with Newton's own notes and corrections, and a specially bound copy of the third edition of the *Principia*, one of only twelve made for Newton himself. Newton's schoolboy copy of Ovid's *Metamorphoses*, purchased when he was just sixteen, with notes in his juvenile handwriting in the margins, sits alongside portraits, medals, and engravings bearing the image of the great man. Grace even acquired a death mask of Newton, one of only five in existence, which had been owned by Thomas Jefferson. It was all housed in the Babson Institute Library in Wellesley, Massachusetts, and looked after by a full-time librarian. Only partly satisfied by the purchase of books and other objects, Grace went further and acquired the parlor room of the house in which Newton lived for much of his time in London. The lonely room still abides at the Babson Institute, where the curious can come to see the same walls, doors, and shutters that Newton saw.

Grace was the driving force behind the Babsons' collection of Newton materials. Roger's involvement with Newton extended beyond the realm of collecting and into a stranger, more unexpected arena: the patronage of Newton-inspired research. In a curious addendum to Grace's book buying, in 1949, more than forty years after he and Grace had first traveled to Europe and commenced their relationship with Newton, Roger established the Gravity Research Foundation, an institution dedicated to understanding—and combating—the force whose universal extent Newton had first posited. In the wake of World War II, Babson feared that American cities could be subject to prolonged attack during the next conflict. The location for the foundation, in the rural enclave of New Boston, New Hampshire, sixty miles north of Boston, was chosen "in case that city should be bombed in World War III."[22] Babson separated the books referring to gravity from the Newton library that Grace had created and shipped them, along with 200,000 "miscellaneous" books, to New Boston.[23] The idea was to preserve a store of essential knowledge relating to gravity (among other

Roger Babson in front of a portrait of Isaac Newton in the room from Newton's London house that Babson bought and reassembled in Wellesley, Massachusetts. In 1949 Babson founded the Gravity Research Foundation with the goal of eradicating from daily life the pernicious effects of gravity, including gravity-related drownings, sickness, and falls. Reproduced courtesy of the Babson College Archives.

things) that would not be lost should the metropolitan center of Boston be bombed.

Having safely sequestered the Newton books, Babson's apocalyptic concerns gave way to a research agenda for the fledgling foundation dedicated to "the study and harnessing of gravity." Though he framed his project in positive terms, it stemmed from a tragedy of his youth. When Babson was a boy, his oldest sister had drowned in the Annisquam River in Gloucester. Babson drew a surprising conclusion about the cause of her death: "Yes, they say she was 'drowned'...but the fact is that, through temporary paralysis, or some other cause (she was a good swimmer) she was unable to fight Gravity which came up and seized her like a dragon and brought her to the bottom."[24]

A further tragedy, the loss of a grandson to another swimming accident, in Lake Winnipesaukee, New Hampshire, in 1947 brought Babson to the conclusion that more research was desperately needed on a scourge that was responsible for a host of ills. He outlined them in a pamphlet titled "Gravity—Our Enemy No. 1," playing off the FBI's famous list. But it wasn't just drowning that gravity could be blamed for. Death by asphyxiation in fires was another kind of gravity-induced drowning, according to Babson, and broken bones were "directly due to the people's inability to counter act Gravity at a critical moment." Respiratory ailments (such as tuberculosis, from which Babson had suffered severely for many years) were caused by the inhalation of the heavy water- and disease-laden air drawn into valleys and houses according to gravity's inexorable pull.[25]

In this life-or-death fight against gravity, it was incumbent upon everyone to incorporate into daily life such practical antigravity aids as life preservers, electric fans (for moving bad air), elevators, and fire detectors. Aside from these practical tools, Babson believed in the potential of more fundamental research into the nature of gravity to yield methods for controlling, or even negating, gravity. To this end he endorsed a prize for the best essay on the possibilities of discovering a "partial insulator, reflector or absorber of gravity waves." One pet idea involved a perpetual-motion machine that harnessed the power of gravity waves and eliminated the world's dependence on nonrenewable fossil fuels.

In 1960 Babson, now eighty-five, still hoped that it would be possible to transform the small town of New Boston into a hub for gravity research. He arranged a free summer conference on his two great passions, investments and gravity, in an attempt to drum up additional interest. His presence at the conference guaranteed some level of attendance despite a strikingly dissonant program. The "Gravity Day" included talks titled "Gravity and the Universe" (by a Columbia University physicist and winner of the 1960 Gravity Essay prize), "Can There be a Shield for Gravitation?," and "The Birds, the Trees and Gravity," given by a local man. The "Investment Days" immediately following included a talk titled "Post-Election Business Outlook," as well as one posing the intriguing question, "After World War III will China take over?"[26]

The Gravity Research Foundation never developed into a center for gravity research, but it went on to play a surprising role in the history of physics. During the 1960s and 1970s, when there was little funding for research on gravity, the Foundation served as a beacon to lonely researchers and a welcome source of support. No less eminent a physicist than Stephen Hawking won the Foundation's award (from first to fifth place) a total of five times between 1965 and 1974. Hawking's now famous pronouncement that "black holes are not black" was first made not in the official article later published in *Nature* but in his 1974 Gravity Research Foundation essay (for which he won third prize). When a colleague commended Hawking for submitting an entry and thereby raising the status of the award, Hawking replied, "I don't know about the prestige, but the money's very welcome."[27] The Foundation continues to makes its presence known today by granting up to five prizes a year for the best short essays on gravity, with awards of up to $4,000.

The legacy of the Gravity Research Foundation extends into granite as well. In the 1960s Babson placed stone monuments—the equivalent of his Dogtown boulders—on the campuses of more than a dozen American college campuses as a reminder to students that the Foundation awaited their work on gravity and antigravity. These monuments remain, reminding students at such institutions as Eastern Nazarene College in Quincy, Massachusetts, and Middlebury College in Vermont "of the blessings forthcoming when science determines what gravity is, how it works, and how it may be controlled" when "a semi-insulator is discovered in order to harness gravity as a free power and reduce airplane accidents."

The stones have outlasted the grand ambitions of the Gravity Research Foundation, though the field of gravity research itself has made strange and wonderful progress in directions Babson never imagined. Grace Babson's Newton collection has similarly traveled in surprising ways. In 1994 the collection was housed with the Burndy Library at the Dibner Institute for the History of Science in Cambridge, Massachusetts. The transfer of the Burndy Library to the Huntington Library in San Marino, California, in 2006 brought the Babson collection to yet another new home. There the collection remains—for now.

10

# The Dealers

From his location in America, A. S. W. Rosenbach traded freely. Backed up by the deep pockets of Grace Babson, he helped to bring Newtonian rarities to a new home across the Atlantic. In London, where the Sotheby sale of the Newton papers took place, circumstances were different. There a group of long-established book dealers participated in a more settled, communal culture of buying and selling. The age of bibliomania in Britain, when aristocratic collectors, book dealers, and librarians rubbed shoulders at the sales, was over. By the 1930s the mania had long subsided into a more structured set of institutions and practices developed for managing the book trade. Auction houses no longer customarily sold directly to private buyers. A relatively small group of book dealers, some of whom represented family-run businesses that had been operating for generations, dominated the auctions, buying lots and later selling them on to private clients. This was a world of pleasantly crammed shops located in Mayfair and Charing Cross. These men were professionals who made their living on the basis of the decisions they made in the auction room, but they were also amateurs in the sense that they loved the objects of their professional attention. "We have chosen to be booksellers not because we believe it is an easy way to make money," explained one longtime dealer, Ernest Weil, "but because either we are addicted to books and reading, and it is our life to us, the real contents of life, or because it is almost the only profession still full of adventures, adventures of the spirit in a world overburdened with routine work."[1]

This combination of vocational clubbishness and avocational passion made English book dealers reluctant to share the auction table with private collectors. The dealers who assembled at Sotheby's on that July afternoon in 1936 were well-known to each other and accustomed to having the room to themselves, literally and metaphorically. By maintaining book auctions as club events, accessible only to members, booksellers were able to keep prices low and their expertise well-guarded until it could be usefully deployed on their own behalf. This required professional discretion and the tacit agreement of those attending to keep bidding moderate and gentlemanly. In reality the lure of a rare item was too great for such informal measures to succeed with any reliability. The truth was that so-called price stability was achieved by what was, by 1936, an illegal form of collusion called a dealer's ring.[2]

Within their covert circles, rings formalized the tacit agreement among dealers to refrain from bidding against each other, keeping prices artificially low. After the auction the whole group customarily retired to the nearest pub, where they held a second auction, called a "knock-out" or "settlement," where the same books were bid for until they reached an authentic market price.[3] The members of the ring then pooled the proceeds from the knock-out sale and divided them equally among themselves. The dealer with the winning bid on any item then sold it to a private buyer at the market rate. In this way the ring essentially ran its own auctions, keeping the proceeds of a sale for its own members, while both the original owner and the official auction house lost out. Dealers who did not actually bid on any items would take away a large (tax-free) sum in exchange for their participation and discretion. One infamous ring that operated at an auction in 1919 at Ruxley Lodge, a baronial estate in Surrey in the south of England, included no fewer than eighty-one book dealers and netted nearly £20,000. Everyone who was anyone in book dealing in this period—Quaritch, Maggs Bros., Tregaskis, and Pickering & Chatto, among others—participated in this ring and no doubt other rings as well.[4]

Though rings were formally made illegal in England in 1927, the passage of legislation did not have much effect. The sale of country house libraries was, according to one London book dealer, "totally dominated by the ring," and it was not uncommon for the auctioneer responsible for running the official auction to hold the second, knock-out sale as well, receiving £10 or so for his trouble. Either these men were shameless, or they did not feel they had much to be ashamed of. It is possible—just—to give a Robin Hood gloss to these operations. That the participants operated according to a sense of redistributive justice (echoed in Munby's reminiscences about freeing up "education plant" for those who might appreciate it more) is corroborated by just what an open secret these rings were. The book dealers found themselves between the sellers, who, even in the throes of massive debt and dispersal, still belonged to a vastly different (and far wealthier) social class, and the auction houses, which were, by definition, larger and more robust than individual dealers. They were also bound by a shared sense that their expertise—as well as the books themselves—was undervalued by owners who had inherited the possessions, but not the passions, of their forebears.

That there was a market at all for the history of science, and the expertise in the book trade to support it, was largely indebted to a small group of Jewish émigré booksellers, many of whom had fled Nazi Germany. Between them, this small group of dealers helped to create an appetite for science book collecting where there had been none before. Even as indisputably important a book as the first edition of the *Principia* could be bought extremely cheaply around 1900. Fifty years later it had become the rare and valuable book it remains today.

First, there was a catalogue. Not since Halliwell had tried to introduce the bibliography of history of science to England had there been anything like a systematic attempt to value books in that field. Before collecting took off, some reference guide was needed to inform potential buyers about what they should be interested in and to provide an indication of what things were worth. Into this vacuum strode Heinrich Zeitlinger, who came to England in 1894 from Linz and quickly found employment with the old book dealing firm of Sotheran's. Zeitlinger was single-handedly responsible for the *Bibliotheca Chemico-Mathematica*, a gargantuan catalogue of "works in many tongues on exact and applied sciences," the first edition of which he started in 1906 and completed fifteen years later, in 1921. The 964-page, two-volume edition included 246 portraits and facsimiles and prices for all manner of bibliographic articles, from a rare first edition of Newton's *Principia* for £18 18*s.* (£400 today) and just £1 15*s.* (just £37 today) for a copy of the second to Tycho Brahe's 1666 *Historia Coelestis* for £2 10*s.*[5]

To read this catalogue today is to disappear down a rabbit hole. Prices and rarity do not seem consonant in this compendium of any- and everything under the sun. An "excessively rare" copy of the 1632 first edition of Galileo's *Dialogue of the two world systems* is listed for £3 13*s.* 6*p* (£77 today), while Copernicus's 1543 first-edition *De Revolutionibus* is listed at what looks, in comparison, like the princely sum of £21 (£445 today). The catalogue contains, in its first edition, some seventeen thousand distinct entries for books and manuscripts. The majority of the contents are not, of course, the great milestones in the history of science. Instead what Zeitlinger had corralled into readable, accessible form was anything that could roughly be classed as scientific investigation. Dizzying comprehensiveness was the point. And so treatises of every length, date, and nationality on subjects from indigo to indivisibles (method of) and induction coils (to name just a sliver of the *I*'s) came together in these crimson volumes.

Once begun, there was no stopping. A separate and new "second" Supplement appeared in 1932. In his preface, Zeitlinger explained that a new edition of the catalogue was needed not simply to include new books but because the prices quoted in the original 1921 edition had been rendered obsolete by rapid rises. In 1932 the first edition *Principia* is not available, while a second edition of 1713, listed as a "large and very clean copy," and also "rare," is priced at £6 6*s*. Galileo's 1632 *Dialogue* has gone up to £35 (nearly £1,200 today), while Copernicus's 1543 *De Revolutionibus* ("excessively rare") has skyrocketed to £75 (£2,500 today). These rises are more remarkable considering the deflation of the British pound in the years since Zeitlinger's first catalogue.

Henry Sotheran, Zeitlinger's employer, boasted that the production was nothing less than "perhaps, the first Historical Catalogue of Science published in any country."[6] He was right. The catalogue spanned all humanity. It was a reference to study and to live with, where what seemed like all of mankind's scientific effusions were made manageable and telling.

Here also was a catalogue that matched a vision of all that the world of science might possibly be considered to contain (Zeitlinger was extremely catholic in his precepts; his topics ranged from the farthest flights of alchemical theory to the most mathematical of treatises) with what the bookish products of those investigations were worth. In doing this, Zeitlinger was performing an act of translation. Why should some bit of scientific arcana be worth so many shillings? The market for these objects was neither large nor coherent. Many of the things Zeitlinger included in his catalogue would previously have been considered part of occult learning or ephemera beneath consideration. By bringing all these wondrous pieces of knowledge together and assigning each and every one a putative value, Zeitlinger was imagining a world in which the material artifacts of knowledge could be assigned a monetary value. These books were valuable not primarily because of who had owned them or what condition they were in (though there was a case to be made for that as well), but because of the historical significance of what they contained. For a certain group of collectors, the history of ideas would henceforth be as desirable as rare literary and illustrated editions.

Compiled as the Great War ripped apart Europe and the technology of mustard gas and automatic weaponry annihilated its soldiers, the catalogue presented a colorful array of the world's peoples united—between its pages at least—in the pursuit of pure knowledge. In the preface Sotheran declaimed, "The pioneers of science have never been of the dryasdust order, and still less confined to one class, but have ranged from the ancients, the Arabs and the great company of the Medieval and after-Reformation clergy, with the remarkable contingent of the Jesuits, to the Lord Mayor of London who first Englished Euclid.... And how many—why not acknowledge it?—are of our own race!"[7] There was room for open-spirited embrace of difference, but the old habits of ranking were not abandoned.

Such a catalogue and the books it contained opened the way into the history of science for those with an interest but no special training, including Professor E. N. da C. Andrade, a physicist who would later chair a Royal Society library committee in charge of editing Newton's correspondence. Much earlier, as a young man in his twenties, Andrade had discovered the thrill of collecting books from the history of science in the decade that Zeitlinger's catalogue had inaugurated. He vividly remembered those halcyon days in an interview forty years later:

> Oh, in '21 I think I began collecting old books, anyhow. I remember going to Edinburgh—in '21 the British Association was meeting there—and I went into Thin's bookshop and asked if they had any old scientific books. They said they didn't think so, and then a youngster said, "Oh, yes, in the cellar we've got the remnants of Professor Crystal's collection." I went down and I spent a hell of a lot of money, I think about 14 pounds [almost £300 today], and I was buying Robert Boyle at about ten bob each [£10 today]. I bought Huyghens' *Horologium Oscillatorium* for 18 shillings [£20 today]. As far as I remember Crystal had given two bob for it [£2 today]. That's a sport that's gone now.[8]

This was the world that Zeitlinger inaugurated with his great catalogues, a world in which collecting rare books seemed as easy as shooting fish in a barrel.

Once started, Zeitlinger was indefatigable. Resuming his trek up the alphabet ladder, he published a second supplement in 1937, fuller and more comprehensive than the first edition (it was just short of 1,400 pages long and contained a staggering twenty-three thousand items). The value of a first edition (and first issue) of Newton's *Principia* had continued to rise, from £18 18*s.* to £42 (£1,500 today). A second but still rare edition (1566) of Copernicus's *De Revolutionibus* could, however, still be had for £9 9*s.* (£350 today), evidence, perhaps, of the supremacy accorded first editions. The autograph of a famous man still claimed a disproportionate price. By 1937 the catalogue included several books that Newton had owned and that contained his signature, including a 1608 edition of Trelcatius ("of great interest as being probably the earliest autograph of Newton"), offered for a not insignificant £10 10*s.* (nearly £400 today), and a presentation copy of Boyle's *Some considerations of the reconcileableness of reason and religion* for the same price.

Prices were only starting to rise in the 1930s. More was to come. The third supplement, issued in 1952, was a document of this transformation. Together, wrote Zeitlinger in the preface, the three editions of the *Bibliotheca* "form a record of the growing scarcity and the still rapidly rising prices of scientific books, which set in some thirty years ago and shows no signs of abating. At one time the cinderella of the book collector, they now appear to have become one of his principal subjects." Books such as Hunter's *Treatise on the Venereal Disease* (1786), Malthus's *Principle of Population* (1798), Boyle's *Sceptical Chymist* (the second edition of 1680), and Newton's *Opticks* (1704) had seen impressive rises in price. As the supply of books from country house libraries dried up, it was good for business, and good for collectors' hearts, to open up the field into "unfamiliar but well-stocked territories." At the same time, as the fortunes of scientific books grew, their rarity increased as they disappeared into private libraries.[9] Books easily purchased at the turn of the century had "practically disappeared" from the market.[10] Owing to the scarcity that Zeitlinger's catalogues had helped foster, many first editions (such as Copernicus's *De Revolutionibus* and Galileo's *Dialogue*) were not available in the third supplement. A first-edition *Principia* was available, however, for £175 (£4,000 today), nearly ten times what it had been worth in 1921.

Coincident with the rise of collecting in the history of science was the rise in lamenting over lost opportunities to buy the great books at scandalously low prices. High prices were the sign of a healthy market for books that had only recently been largely invisible to collectors. For bookmen, recognizing the value of objects from the past was a matter not of sentiment nor theory but of business. For those with the foresight to appreciate the books before they became too rare or expensive, great deals could be had and good money could be made. As far back as 1862 book buying had been recognized as a species of prognostication. "It is the general ambition of the class to find value where there seems to be none," wrote John Hill Burton in *The Book Hunter*, to look into a "heap of rubbish" and to put his finger on something "valuable and curious."[11]

That intuitive, prognosticating skill elevated the bookman in the eyes of scholars. The distinction between dealers and scholars had been murky in the early days. As both history and book dealing firmed up their professional boundaries (and their codes of practice), the connoisseurial skill of the book dealer remained important, despite (or perhaps because of) its often eccentric nature. The best booksellers were also scholars who harbored a taste for the books they sold. Though professional scholars tended to belittle the mercantile associations of the bookseller, the seller often saw ahead of the scholar. It was still one of the collector's "most significant functions to anticipate the scholar and historian," wrote collector and dealer John Carter, "to find some interest where none was recognized before, to rescue books from obscurity, to pioneer a subject or an author by seeking out and assembling the raw material for study, in whatever its printed form."[12]

This might explain the preponderance of Jewish book dealers in the early days of collecting in scientific books. Booksellers such as Heinrich Zeitlinger, Ernst Weil, and E. P. Goldschmidt had left their homes in Germany, Austria, and Belgium to travel to England. Once there, they remained outsiders by dint of their religion, their nationality, and what was referred to as their race. There were disadvantages to being an outsider, even in England, but alienation also had its rewards. In reckoning the scientific past, the backward gaze of an outsider was useful. The émigré discerned value and meaning where the native might not. More concretely many of the book dealers had brought their stock with them, physically transporting some of the contents of their catalogues to the shores of England. Dating back to the great sales of Libri's ill-gotten gains in the 1860s, many great European books and manuscripts had made their way to England. Zeitlinger's catalogue was by no means a purely English one, despite Sotheran's brag.

Ernst Weil offers a good example of how Continental interests—and stock—came to the English book trade. Born in 1891 in Ulm on the Danube, Weil had come to England to learn banking before returning to fight on behalf of Germany in the Great War. He attained a doctorate in art history before starting an antiquarian bookshop in 1923 with I. W. Taeuber. Their bookshop was one of the first to specialize in early medicine and the natural sciences. (According to Weil's daughter, Ernst had a soft spot for alchemy.) Unfortunately Weil's partner himself preferred National Socialism, though the men somehow remained friends, an indication, perhaps, of the deep bonds that unite bookmen.

But even those ties were not strong enough in the face of the incoming tide. The bookshop was located opposite Hitler's Munich headquarters. As early as 1932 there were SA uniforms hanging in the entrance hall when Weil arrived for work. It was time to leave. Weil and Taeuber parted on good terms, dividing up the books equally. Weil and his family left for England in June 1933, when, his daughter reports, "it was still possible to take all your belongings with you." So some of the Weil stock ended up in London, more specifically, with the firm of E. P. Goldschmidt, where Weil worked for ten years before he and Goldschmidt parted ways.

Goldschmidt had also lived in England before the Great War, studying at Cambridge in the early years of the century, when Keynes was there and when the Goldschmidt family fortune was still large enough that he was known as the richest undergraduate in attendance. He had fed his appetite for books with purchases from Gustave David's bookstall in the market square. Inflation and taxes following the war decimated his inheritance, and when he returned to England in 1923 it was with vastly diminished funds. His passion for books was intact, however, and he set up shop on Old Bond Street. He smoked a distinctive blend of tobacco, and it was said the provenance of more than one book had been settled by simply opening it to allow the woodsy aroma to be inhaled from its pages.

Goldschmidt's daily routine was the stuff of legend. Weil reported that he did not usually see his employer until 1 p.m., when they promptly went out to lunch. This was followed by a bit of "playful fight" with the cat, a stroll, and some research in the shop, where the final business for the day was resolved. The *Times* crossword puzzle was solved before dinner, and following dinner, a visitor often arrived. Only "well after midnight," when everyone else had retired for the day, did Goldschmidt sit down to compose his famous book descriptions on the little square slips of paper he preferred to use. "Before retiring in the early hours of the morning," Weil concluded, he often went to the all-night Lyons Corner House near Piccadilly for a glass of milk.[13]

Goldschmidt was the consummate scholar-bookseller, who, despite his shrewd eye for business, was dismissive of the financial aspects of the trade. The ideal customer, he liked to say, was "a man who lived 2000 miles away and occasionally ordered by postcard a very expensive book."[14] He was unstinting in his disdain for the vagaries of fashion in book collecting, such as the fetish for illuminated books and manuscripts—the pretty dimwits of the trade. He noted that most libraries refused to acquire non-illustrated manuscripts unless burdened with them as a gift. Despite his love for books, Goldschmidt was clear-eyed about what (most of) the rest of the culture thought of them. Even though books, as the material artifacts of knowledge, preserved the invaluable history of how that knowledge had been acquired, their low value was no anomaly but a "structural feature" of the very society that had benefited from the knowledge they contained.[15]

Goldschmidt was right. The few books that did emerge from Europe were haunted by the many more that didn't. Walter Mehring's 1951 *The Lost Library: The Autobiography of a Culture*, is just one poignant example of the destruction of libraries (in this case his father's) at the hands of the Nazis. Books were the concrete emblems of a culture and, like the human victims of the war, mobile signs of both the tenacity and the vulnerability of that culture.

In America there was a taste for list making rather than elegy. In 1934 an exhibition was mounted in Berkeley to commemorate 114 "first editions of epochal achievements in the history of science." Here was a sharp contrast to the wild effusion of human knowledge in Zeitlinger's *Bibliotheca*. In exchange for the tangled bank of learning, Herbert McLean Evans, the anatomist and embryologist who curated the exhibit, offered a league table of discoveries, laws, and hypotheses that had among them been "directly or indirectly responsible for the advancement of science." Evans's justification for the backward gaze in science was tellingly explicit: there was still much to explain to a wider audience (the scene was the ninety-fourth meeting of the American Association for the Advancement of Science) about why science in the past mattered.

However, the traditional pleasures of book collecting were still alive, particularly the collecting of first editions, which was one "of the chief cults of bibliomania," and applied more to science than other areas, argued Evans. While literary editions preserved and repeated the original text of the first edition (more or less), in science the past was by definition fleeting and quickly abandoned. This rush toward the future made the preservation of the record of past achievement all the more important. Evans referred to George Sarton, the father of academic history of science in the United States, who asserted that "knowledge as opposed to beauty, is cumulative and progressive." Only by looking at the original idea could you hope to "observe the origin and change of ideas."[16] The first edition in science offered not only the opportunity to assess the past but to gauge how far we'd come.

Evans's exhibit was commemorated in a tiny little pamphlet that is easy to miss on a bookshelf surrounded by more capacious catalogues like the *Bibliotheca.* By paring the field down to a list of greatest hits, Evans emphasized the progression of ideas in an orderly lineage. This was a more selective and ultimately straightforward vision of the history of science than that forwarded by Weil, with his secret delight in alchemy. Evans's delight in the past was in the excavation of the shining path that led to those very contemporary accomplishments. His list may have been short, but it became just as influential as Zeitlinger's hefty tomes. Evans's approach provided readers with a handy guide through the thickets of the past. Such lists proved especially desirable to the wealthiest of collectors who could still afford to buy the great books they enumerated. For this reason, many of the choicest books on Evans's list soon become unaffordable to most buyers, and in the dark days of the Depression, history of science books held their value better than any other books.[17] Those few who had enjoyed the field in its infancy, nostalgic for the bygone bargains of the early decades of the century, were not always pleased by the rise in prices.

For book dealers who had to earn their bread and butter through the days and weeks, there were other ways to navigate the byways of the trade. Buyers were grateful for pointers about ways of escaping "from the expanding pressures and spiraling prices of an overcrowded and ever more shrilly exploited market."[18] Science was, after all, much more than a series of thunderclaps signaling revolutionary discoveries; it was more like a constant patter of documentary rain. "In science," explained Weil, "the great events occur with long intervals between them, not empty intervals—they are filled with smaller work, with books, with papers which prepare them." There was much to collect between the "milestones of civilization."[19] Just as the high points of science were separated by stretches of preparatory work, so the book trade must learn to supplement their first-edition feast days with everyday trade in less revolutionary material. Weil described how he had contacted Roentgen's publisher to hunt down any surviving offprints of his 1896 paper on the discovery of X-rays. The key was to find the material before it was destroyed and before anyone else got there first (or thought to do so).

The Newton papers on Sotheby's auction block were overwhelmingly examples of what, at best, Weil would have described as work "preparatory" to the so-called milestones of civilization. In fact Newton's "foul" and reworked notes were more challenging still because their subject matter—theology, chronology, alchemy—stretched to its limit a nascent definition of what counted as the history of science. Given their ambiguous nature, it was not at all clear how the sale at Sotheby's would go.

11

# The Sotheby Sale

On the day of the Portsmouth family's sale of the Newton papers another sale was happening less than a mile away from Sotheby's, on the other side of Mayfair. In their set of rooms on St. James Square, Christie's—Sotheby's historic rival—was overseeing what would turn out to be the biggest auction of art and collectibles in London for at least a decade, a fifteen-day extravaganza of objects from the collection of Henry Oppenheimer, including antiquities, majolica, Renaissance medals, and the best set of Old Master drawings then held in private hands. The Oppenheimer sale was a chance for London to prove its primacy on the international scene and for private buyers from around the world to snap up masterpieces that, if not exactly bargains, would perhaps not be seen again on the open market for years to come. It wasn't a matter of making easy money but of investing for the future. Drawings by Old Masters had already proven to be art-world winners, exquisite in their beauty and highly commendable in their tendency to accrue value as the years went by. That was more than could be said for most investments at the time.

When the Sotheby's auction of Newton papers opened, the Oppenheimer sale at Christie's was already in its second week, having smashed auction house records by taking in more than £30,000 on the first day of bidding. (By the end of its two-week run, it would net more than £140,000, more than £5 million today.) The timing of this lesser sale may have been inopportune (though then, as now, the buyers for old manuscripts belonged to a different set than those for old art), but it had been planned and executed in haste. Sotheby's had mounted the auction at Wallop's urgent request and scrambled to produce a sales catalogue for the complex archive in just under four months.

The need for catalogue copy in advance of the July auction date put James Cameron Taylor, the man assigned the task, under considerable pressure to work quickly and accurately. Taylor had left school at fourteen but studied classics while working at Hodgson & Co., a well-established bookselling firm on Chancery Lane. Wounded in the trenches in World War I, he then worked on a secret project to produce nitroglycerine for the Ministry of Munitions. By 1936 he had been at Sotheby's (aside from the wartime hiatus) for nearly twenty years. Perhaps unsurprisingly, given his evident strength of character and nerve, Taylor accomplished the Newton catalogue on time and with such accuracy and depth that it remains an important document of reference even some eighty years later.[1]

The catalogue revealed the fact, based on estimates by Taylor, that the papers in the sale contained more than three million of Newton's own handwritten words. Even for a novice, this was an impressive figure. It was perhaps even more noteworthy that some 1,250,000 of them, more than a third, were on theological subjects, while more than 650,000 were on alchemy and nearly 250,000 were on chronology. Beyond that, the richness of the archive was conveyed to the public through biographical anecdotes: how the infant Newton had fit into a quart pot; the ink bottle that the student Newton bought on arriving in Cambridge in 1661. The treasures of the sale included letters as well as manuscripts, including "Boyle writing to tell Newton of the appearance of a comet, Locke sending him a recipe for making gold, and Pepys beseeching him to put his mathematics to some really practical use by computing for him the chances of throwing sixes at dice."[2]

Taylor had done a superb job of identifying and contextualizing the range and richness of the material, startlingly so given the scant time he had in which to do the work and the breadth (rather than depth) of his own knowledge. But the catalogue could go only so far in transforming an archive that had stymied the most energetic and committed of men for two centuries into something that could be valued by the marketplace. The papers, covering theology, alchemy, and administration of the Mint, were not scientific in the obvious sense, but neither were they literary. There were few, if any, scholars for whom such material was of immediate interest and fewer still with the historical skills to fully digest them. They represented aspects of Newton's thoughts that were generally unstudied, which did not give them a ready market. For dealers and collectors, the Newton material was primarily of interest as autograph material: writing from the hand of Newton.

This would tend to suggest that a ring did not control the auction. There had been enough presale publicity—the Newton name was hard to miss—that anyone with an interest would have known about the sale and been well apprised of the contents. No shady descriptions such as "loose bundle of papers" showed up in Taylor's catalogue to help a would-be ring obscure their takings. And the dealers were not alone at the sale, as might have been the case for a ring event. Wallop, for one, was there. And Keynes, whose fame extended well beyond the esoteric realms of book collecting. Keynes was undoubtedly respected by the book dealers in the room as a collector of serious and informed intent with the pocketbook to match his ambitions (in other words, as the best sort of customer they could wish to have). It would have been foolish, and foolhardy, to con the great economist. He, more than anyone else, was prepared to see the value in the strange papers on sale that day.

Keynes had begun collecting books as a schoolboy. He had trained his eye at Eton and during holidays at home in Cambridge buying up Aldines and Elzevirs, highly collectible early sixteenth-century editions of mostly Greek and Latin texts. Keynes was still a boy when he made the acquaintance of Gustave David, who had a bookstall in the Cambridge marketplace and inculcated in a cadre of young men the addictive pleasures of book collecting. David was a Cambridge eminence, a "Bowler-hatted, chain-smoking old bouquiniste who had migrated from the banks of the Seine to Cambridge Market Place."[3] A contemporary photograph shows him in situ, sitting formally under a dingy awning next to an array of books set on a table, spines upward for easy browsing. The damp, gray air of the marketplace is almost palpable, as is the spirit of enduring custodianship emanating from David.

The book dealer didn't spend all his time at the market stall. He spent Thursdays in the London salerooms buying job lots, miscellaneous bundles of books in which treasures might be hidden, and returning to Cambridge on Fridays to bestow his best finds on his best young men. David had a destabilizing way about him and was "something of a tyrant and awarded his books rather than sold them—and awarded them in a most arbitrary and unpredictable way," reported Munby, who had himself bought books from him.[4] Keynes became a favorite of the bookseller before he joined that other group of serious, self-regarding readers, the secret intellectual society of the Cambridge Apostles. Just as those Apostles numbered

Gustave David at his book stall in the Cambridge market square. He sold Keynes a first-edition *Principia* for 4 shillings, having paid 4 pence for it himself. Courtesy of G. David, Bookseller.

their members in order of acceptance (Keynes was 243), Keynes numbered in order of acquisition the 329 books he purchased as a schoolboy from David.[5]

Keynes's fascination with books only grew. The exhilaration of starting university was tied up with book buying. "I know nice people," he wrote to a friend his first month at King's. "I have bought over fifty books this term. I row hard every afternoon without exception, and I never go to bed. What more can heaven offer me?"[6] A first edition of the *Principia*, perhaps? Still in Keynes's library, the inscription (in one of four editions that Keynes acquired) attests to the delights of the time: "I purchased this from David about 1905 for four shillings, he having bought it in Farringdon Road for four pence."[7] Keynes continued collecting throughout his busy, increasingly public life, as he combined donnish pursuits with high-level government advising. Well into the early 1930s David continued to set aside books for him.

And so it went, until right around the time of the Sotheby sale, when something shifted in Keynes and he began to collect with a new forcefulness and a new direction. Henceforth until his early death, some seven years later, at just sixty-two, he would largely abandon literature and be steadfast in buying first and early editions of European philosophy, political economy, and science.

Blessed with the means to purchase more or less what he wished (by this time he had accumulated substantial wealth as a result of astute investments), and with the discipline to buy in set areas, Keynes was a passionate, informed, and canny collector. He was also, rather unusually for a collector, a voracious reader. He found pleasure in his books from the moment he began his perusal of the sales catalogues through the thrill of the chase and purchase, and finally to the hours he spent with them in his library at Tilton or in London. He was both systematic and energetic. He read and annotated catalogues and sent commissions, or advance bids, to nearly all of the important auction sales. He tried to see the books he aimed to purchase, unless he was really busy (which was a very high bar for Keynes). As a result, he got first and early

John Maynard Keynes was already an avid book collector and reader in 1908, when this portrait was made. © National Portrait Gallery, London.

editions of an impressive number of the greats he was after: Bacon, Bentham, Berkeley, Copernicus, Descartes, Hegel, Hobbes, Hume, Kant, Leibnitz, Locke, Malebranche, Malthus, Newton, Pascal, Rousseau, Adam Smith, and Spinoza. (His collection of Newton's printed editions includes, in addition to the four first editions of the *Principia* [1687] mentioned above, three second editions printed in 1713 and four third editions from 1726.[8])

It is worth considering this list for a moment. These names, familiar and imposing figures in the history of Western thought, had never before found much favor on the shelves of book collectors. It was the art of printing itself that first captured the imagination of most collectors: illuminations and illustrations, fine bindings, and rare examples of early presses (those Aldines and Elzevirs). Literature, poetry, and drama, especially from the Elizabethan and Stuart periods, was also deeply fashionable. By Keynes's day the first edition had become the paragon of desirability. The lust for the perfect edition—uncut pages, clean boards, the absolute minimum of wear—meant that a history of use was anathema. Ownership rather than readership was the fetish of collectors. The history of a book's possessors, the more refined the better, added value to a book, but reading the books that one collected, and caring who might have done so in the past, was largely an irrelevance to collectors at the time. Reading could be done, but it was not necessary and, in the case of the uncut pages, inadvisable.

Keynes's new style of collecting was self-consciously intellectual, as opposed to aesthetic or literary. It asserted that a particular history of ideas or chain of thought linked certain men through the ages. And it projected the implicit assumption that its creator was an inheritor of both the material and the intellectual masterpieces of a previous age. Keynes was a thoroughgoing Bloomsburyite in this respect. The paintings on the wall, the rugs on the floor, the furnishings in the room, and the books on the shelves were never just things: they were the physical embodiment of ideas and values whose display was a source of both aesthetic pleasure and moral reinforcement. A book in the hand, like the good life in Bloomsbury or the Sussex countryside, linked the life of the mind with that of the physical world.

This change in the focus of his collecting may also have reflected a more personal aspect to Keynes, a man who shaped policy at the highest level in Britain while privately pursuing more personal forms of truth in history, philosophy, archaeology, and the arts. Keynes had long been famous as a result of his role in shaping British financing during the Great War, his prominence as a writer of essays and reviews for the press, and his best-selling tirade against postwar reparations, *The Economic Consequences of the Peace.* In 1936 his fame had dramatically increased with the publication of *The General Theory of Employment, Interest and Money,* in which he detailed policy for optimizing employment and mitigating inflation. At the same time, Keynes was a person who led what his biographer Robert Skidelsky has called a "coded" life, with much that was essential hidden from prying eyes. This instinct for privacy extended to the belief that "beneath the knowledge in which he publicly dealt there lay an esoteric knowledge open only to a few initiates."[9] Within the intimate confines of the books of great philosophers, Keynes may have felt he could safely enter into a deeper, truer world of knowledge. There could be no better preparation for an encounter with the Newton papers being sold at Sotheby's than the belief that beneath the veil of the everyday lay hidden truths, waiting to be uncovered by a select few.

Fittingly Keynes himself provides the best description—and implicit defense—of the world of books, a world in which mystery could be welcomed merely by crossing the threshold of a bookshop. In an address he delivered on the BBC less than two weeks before the Newton sale, he shared "a little general advice from one who can claim to be an experienced reader" with "those who have learnt to read but have not yet gained experience." It is vintage Keynes, kindly and superior, possessed of equal measures of intuition and skill, at home in a world both sensuous and profoundly intellectual. A reader, he advised, should approach books with "all his senses," touching them, rustling their pages, and even smelling them. He should live surrounded by books, with a "penumbra of unread pages" around him. The sense of mystery in the fluttering pages, of pleasures unknown waiting to be fulfilled, was the purpose of libraries and, especially, good bookshops. "A bookshop is not like a railway booking-office which one approaches knowing what one wants," advised Keynes. "One should enter it vaguely, almost in a dream, and allow what is there freely to attract and influence the eye. To walk the rounds of the bookshops, dipping in as curiosity dictates, should be an afternoon's entertainment. Feel no shyness or compunction in taking it. Bookshops exist to provide it; and the booksellers welcome it, knowing how it will end."[10]

John Maynard Keynes as he looked around the time of the Sotheby sale of Newton papers in July 1936. He published his landmark work, *The General Theory of Employment, Interest and Money*, just a few months earlier. © National Portrait Gallery, London.

By the time he turned up at the Sotheby's sale that July afternoon, Keynes was primed to recognize that the papers were valuable not simply as material artifacts but as the record of a great man's thought process. He was also prepared to spend the cash needed to secure them. In the early years of the decade, his investments in commodities and currencies had been spectacularly successful and his net worth peaked the year of the sale at around £500,000 (almost £13 million today). Other buyers included Maggs Bros. and Bernard Quaritch, the preeminent dealers in rare books and manuscripts of the period whose firms dated from the mid-nineteenth century; E. P. Goldschmidt; a representative from the Cambridge dealer Heffer and Sons; Gabriel Wells, a Hungarian who had emigrated to New York; and Emmanuel Fabius, the only French dealer to attend. Maggs Bros., which benefited from extended credit from Sotheby's, was the most aggressive, and successful, bidder at the auction, taking away eighty-nine lots in total, while

Heffers and Wells ended up with twenty-four and twenty-three, respectively. Fabius acquired thirteen, while a pseudonymous "Ulysses" got sixteen. Francis Edwards, another London dealer, bought twelve, and Gerard Wallop bought back ten lots.[11]

Keynes was the second most successful buyer, with a total of thirty-eight lots, mostly items relating to Newton's alchemical work. He bought thirteen lots on the first day, for £264 10*s.* (nearly £10,000 today). His acquisitive appetite whetted, he returned the next day and stepped up his bidding considerably, buying twenty-five items for a total of £391 (just under £15,000 today).[12] Though he bid confidently, even aggressively on some items, he let others go. His desire was tempered by discretion. He was savvy enough to understand the rules of the game. Even without a formal dealers' ring in operation, he knew that it would be unwise to create a bidding war in what amounted to a closed auction: aside from Keynes and Wallop, only professional booksellers were present at the sale.

All told, there were thirty-seven purchasers at the sale, of whom nine bought ten or more lots.[13] And so the bulk of Newton's remaining

The room at Sotheby's in which the Newton papers were sold. The occasion is the 1947 sale of C. W. Dyson Perrins's Gutenberg Bible. Ernest Maggs, of Maggs Bros., is the man with the goatee second from the auctioneer on the left side of the table. Courtesy of Maggs Bros. Ltd.

manuscripts, kept intact and relatively safe for more than two hundred years following his death (many of those dating from his early life were more than 250 years old), were scattered to dozens of buyers all over the world, some never to be seen again. The final hammer price was £9,000 (£330,000 today) over the two days. Certainly when compared with the £140,000 raised by the Oppenheimer sale, the Portsmouth Sale was disappointing. But from the perspective of Keynes, the buyer, it had all gone "extraordinarily reasonably" (though Keynes thought it "rather peculiar" that neither the British Museum, the Cambridge University Library, nor Trinity College, Cambridge had even sent representatives to the sale).[14]

For some observers, the dispersion seemed a tragic loss of patrimony. While fashions were fashions, and manuscripts commodities that followed the rules of the marketplace, Newton was Newton. Surely that must have been worth something, if not to private collectors then to the nation itself? Notwithstanding the more pressing concern of a possible war with Germany, there was some precedent in this period for public bodies to step in and save valuable artifacts for the nation (though legislation formally authorizing government protection of national treasures would not be passed until war broke out in 1939). When a collection of Lord Nelson relics came up at auction in May 1936, just two months before the Portsmouth sale, the National Maritime Museum arranged for them to be bought for the nation, including a silver sauce tureen presented to Nelson by Lloyd's after the battle of Copenhagen in 1801 (which went for £500).[15]

Three years earlier private individuals and groups across the nation had contributed more than half of the £100,000 needed by the nation to purchase the *Codex Sinaiticus* (sold by Maggs Bros. on behalf of the Soviet government, which needed cash for its second Five-Year Plan). The *Codex*, the earliest complete copy of the Christian Bible and an incomparable source for historians of religion, caught the public's imagination. Crowds thronged to see it on display. Librarians assured the public that the vast sums needed to buy the *Codex* were worth it. And the public responded with an outpouring of funds to ensure that it could stay. Aside from its evident appeal as a religious artifact, the *Codex* communicated the superiority—even necessity—of original manuscripts. The subtleties of ink and hue were such that to truly understand the history of the manuscript, and thus the history of the Bible itself, mere copies would not suffice.

Newton's mystical, administrative, and alchemical writings did not have the pull of Nelson's heroics nor an ancient *Codex*. As Keynes noted, no representatives from institutions of higher learning and research such as the British Museum, Cambridge and Oxford universities, or Trinity Library had attended the Newton sale, and there was little fanfare afterward, save for a brief flurry of letters to the editor of the *Times* that suggest a modicum of interest in Britain's finest scientific mind. The *Times Literary Supplement* the week following the sale raised a solitary note of opportunity missed: "One cannot help regretting that it was not possible to buy the entire collection for an English library. Someone, surely, if properly approached, would have been prepared to find, say £10,000 to acquire a valuable body of papers such as this, including some three million words in Newton's own hand."[16]

It had not happened. Yet all was not necessarily lost. Perhaps the manuscripts were not as requisite to a full study of Newton as the *Codex* was for Christianity. From Cambridge, Joseph Larmor, the man who had edited Stokes's scientific papers and correspondence, wrote a letter to the *Times* noting that the dispersal wasn't as significant as some might think. The existence of the Portsmouth catalogue created by Adams, Stokes, Liveing, and Luard, along with the "indispensible account of the astronomical part of the contents of the manuscripts," helped to soften the loss and "in fact reduces their importance mainly to the status of original autographs."[17] Autographs might amuse the connoisseurs who clamored for more Byron, more modern firsts, but in Larmor's view historians of science and astronomers could manage well enough with the Cambridge catalogue.

Others thought differently and set out to gather the papers back together as best they could. If the Sotheby's sale had dispersed the papers with stunning speed, it also presented an opportunity to bring them into the hands of those who could assess their true value. Keynes, for one, was resolved. After the sale he immediately set about buying up lots that in retrospect he realized that he wanted. The closeness of his relationships with the book dealers now came strongly into play, as he started contacting the firms that had bought the lots he was hunting down. The letters reveal how much the professionals trusted and esteemed Keynes. Newton manuscripts started circulating via the Royal Mail as Keynes requested items he had missed at the auction. On August 18 the bookseller Francis Edwards sent four items to Keynes at his country house in Sussex "on approval." Priced at nearly £100 (£4,000 today), the manuscripts were mostly satisfactory to Keynes, who placed three of his customary crisp check marks on the invoice list next to the volumes he wished to keep and returned the fourth with a request for another lot from the Newton sale.[18]

Keynes wasn't the only person dismayed that the Newton papers had been so scattered by the sale. Lord Wakefield, more commonly known as "Cheers," who had made his wealth on the proceeds of industrial and automotive lubricants, had a track record of public-spiritedness. Three years earlier he had anonymously contributed to the *Codex Sinaiticus* appeal and had presented Lord Nelson's personal log book to the nation. Now he spent £1,400 (£50,000 today) to buy Newton's Mint papers, bound in three handsome folios, from Gabriel Wells, who made them available at cost with the understanding that they would be donated to the Royal Mint. And within two weeks of the sale, another donation was made, this time a set of manuscript letters purchased by Sir Robert Hadfield and presented to the Royal Society.

But Keynes had bigger ideas. He wrote to the French dealer Emmanuel Fabius to ask after some lots he had bought containing correspondence between Newton and Bernard de Fontenelle, the secretary of the Académie des sciences. Keynes was willing to pay 1,500 francs for the two lots.[19] Fabius asked for 9,000 francs (more than £120 then and £4,400 today), on account of the "extreme rarity" of Newton manuscripts, particularly those relating to France. Put off, perhaps, by Fabius's price, Keynes turned his search closer to home and began contacting dealers in Cambridge and London who had been at the sale. Keynes's resolve to create a comprehensive collection of Newton material strengthened in the weeks following the sale. Maggs offered Keynes whatever he had that was of interest at just 20 percent commission. Keynes initially spent £42 and wrote to thank the book dealer for the "very moderate" prices. "I have now decided to form a very substantial collection of these papers with the idea of keeping them permanently in Cambridge," he went on to explain, "and with this object in view I have been going through the catalogue carefully."[20] The upshot was that he had found more lots bought by Maggs that he was interested in buying. Keynes was getting in deeper and deeper. Enclosing a further check for £653.15 (£16,000 today) to cover additional purchases from Maggs, he wrote, "I am afraid that my appetite grows on what it feeds on."[21] He asked to buy ten more lots and to see four more on approval.

It didn't hurt to indicate a worthy aim when endeavoring to secure the papers for a good price, and he also mentioned that he was planning to donate the material to Cambridge in letters to both Goldschmidt and Heffer.[22] Heffer's reply captures the mood of the times: "In the ideal world I suppose the MSS of Newton would never have been sold, and in the 'next to the ideal' world I suppose that whoever bought them would have given them to where they belong. But unfortunately this is a commercial world, where the bitter struggle to make both ends meet goes on unceasingly." That commercial world, however, did not extend to the still gentlemanly relationship between a book dealer and his long-term customer: Heffer went on to offer Keynes the very low price of just £6 10*s.* for the lot (number 166, Newton's letters to Roger Cotes, the mathematician who helped Newton prepare the second edition of the *Principia*), only £1 more than he had paid for it.[23] Keynes wrote back immediately to thank him and say that he would be happy to pay what he considered an "extremely moderate" price.[24] Heffer had other Newton items, and while Keynes didn't get everything he wanted (some were out on approval to another prospective buyer), he ended up spending nearly £100 with Heffer.

Keynes also received a letter from Gerard Wallop, asking if he would be interested in selling back Lot 295, a draft of Newton's writing "on the origin of monarchies." Wallop explained that it was only the very heavy death duties following his aunt's death that had forced the Sotheby's sale and that he particularly regretted the loss of Newton's writing with a political slant. He offered a generous £50 for the manuscript (Keynes had paid £11) and invited Keynes to his home for a meal.[25] Keynes replied that he would be happy to arrange for the "Origin" to be sent back to Hampshire and that he would enjoy a visit in the new year.[26]

On September 17 Keynes wrote back in reply to what he called Fabius's "interesting" letter asking for 9,000 francs for the Fontenelle letters and apologetically excused himself from the purchase on account of the high price. Fabius may have been pushing his luck, but it's equally possible that his suggested price reflected something closer to market rates than the cozy 20 percent increases that the London sellers were offering Keynes. In any case, the loss of the Fontenelle letters didn't matter too greatly to Keynes. By then he had already acquired 130 lots from the sale.[27] He had also started studying alchemy in earnest and requested a catalogue on the subject at the end of a letter to Maggs.[28]

Soon Keynes received the first hint that someone else was as interested in buying the sale lots as he was. He had tracked Gabriel Wells to a hotel in Budapest with a request for lots 72, 222, 223, and 320. Wells lamented, "I would your letter had got to me sooner. It so happens that Prof. Yahuda has kept after me all along."[29] Keynes replied to say that he was "sorry, though not surprised" to have missed his chance on some of the lots but that he considered his position a good one: "I have now acquired altogether more than one-third of the total number of lots in the Sale, and, in my own judgement, so far as interest goes, considerably more than half."[30] He also admitted to Wells (after snapping up two more of Newton's manuscripts from him) that it was his fault to have missed out on additional lots. "I only gradually came to the decision to make my collection comprehensive. It is only gradually that I came to the view that the papers concerning alchemy were really interesting."[31]

Soon Keynes and Abraham Yahuda encountered each other more directly. On September 7, 1936, H. Clifford Maggs wrote to Keynes on behalf of Yahuda to say that the latter "would be glad to get into touch with you as he finds he has some material belonging to one of your lots, No. 263, in the Newton sale."[32] Keynes was "very much interested" in this information on a long manuscript relating to the Temple of Solomon. "I now hold the main bulk of the papers," he replied directly to Yahuda, "and would like to have the homes of any other considerable collection of the papers duly recorded."[33]

The Sotheby's cataloguers had divided up an unruly mass of manuscript papers as best they could, but some material had been clumsily separated. The Solomon's Temple manuscript, Yahuda claimed, almost certainly related to other materials on the Temple he had already acquired. Yahuda, it soon became clear, was interested in more than simply returning a mislaid manuscript. He continued to the crux of the matter: "Now as I would like to have all Newton has written about this material together, I thought you might perhaps be persuaded either to sell that lot or to exchange it against other Newton lots on alchemy which I have in my possession." Though their interests and backgrounds were quite different, Keynes and Yahuda voiced the same rationale for their collecting: to keep the papers together. "You were perfectly right to secure a great number of Newton papers so that they should not be dispersed more than they have been," wrote Yahuda to Keynes on September 15. It was the motivation behind his own collecting, as it was for Keynes.[34]

Their desire to keep the papers together did not blind them, however, to the potential advantages to their collections of a well-considered trade. Though Keynes was reluctant, he was not ultimately unwilling to exchange.[35] Luckily the men had different areas of interest in the Newton papers, with Yahuda pursuing theology and Keynes alchemy (though Keynes ended up with some significant theological works and Yahuda with some alchemy). The letters documenting this exchange do not survive, but the respective collections of Keynes and Yahuda based on their trades do—at King's College, Cambridge, and at the National Library of Israel.

Yahuda and Keynes corresponded over a period of two years about the papers and related subjects, such as the preparation of a volume to celebrate the tercentenary of Newton's birth. In a letter of April 3, 1938, Keynes shared his developing thoughts about Newton's alchemical writing: "What is it Newton really thought he was doing? Is it all the usual stuff and nonsense? Or is there some glimmering of the beginnings of real chemistry? I fancy that this could only be answered by someone who was primarily competent on the chemical side and was looking at the stuff as a just possible contribution to science and not merely to dead commentary."[36] At the end of this letter, Keynes inquired after a manuscript he had sent to Yahuda as part of a proposed exchange. Yahuda's reply, that he was unable to find any lots in his possession belonging to Keynes, ends with a hope that the two men can meet to discuss Newton's theological, chronological, and polemical writings, to which he attributed a "much greater importance than has hitherto been done by his critics."[37] Evidently Yahuda had been reading Newton's theological works with a seriousness of purpose that they had not received for some time, if ever. His conclusion, briefly stated in another letter to Keynes, was that "Newton's efforts meant more than a simple aberration in the sterile fields of alchemist adventures of cabbalistic charlatanism."[38] That exchange is the last we have between the two men. The matter of the traded manuscripts was never settled, and perhaps Keynes lost patience with Yahuda. In any case, Keynes had created a collection of which he was proud and with which he was content.

Keynes set about trying to understand more of what the mass of papers contained. This was no easy task. Most of the material consisted of Newton's notes on or copies of other alchemical writings, as well as anonymous treatises in unknown handwriting. It was hard to discern anything empirical, original, or finished in the mixed bag of writings. Anything resembling an experimental notebook on alchemy was conspicuously absent. And yet Newton had evidently taken care to keep these papers throughout his long life. One of the first men Keynes contacted in his quest was L. F. Gilbert, of the Society for the Study of Alchemy and Early Chemistry. Keynes noted in a letter to Gilbert that Newton had deliberately chosen to preserve these papers, even as he burned others. There was something here that Newton thought worth substantial amounts of his time and worth saving for posterity. Yet,

Keynes went on to say, "his selection of the Alchemical MSS [to save] is exceedingly queer, most difficult to attribute to the same brain as that which produced his other works." Keynes had been forced to conclude that Newton had, in fact, been producing the alchemical papers immediately *before* he embarked upon the *Principia*. He was indeed the same man, the same brain, whether he was doing natural philosophy or, as Keynes called it, "this queer stuff."[39]

Gilbert tried to help Keynes reconcile this, while at the same time considering the possibility that the alchemy indicated a deviance that was potentially significant. It *was* hard to attribute "to the same brain his other works and his absorption in the queer stuff," he responded to Keynes, but Newton wasn't the only person who admitted a quotient of oddness into an otherwise reasonable self. This sort of thing was historically consistent. There were, after all, plenty of people who were more or less sane but with "marked kinks when it comes to certain affairs. When those kinks become too objectionable these people have to go to asylums."[40] In a subsequent letter to Keynes, Gilbert qualified his view "about Newton having a kink," which had not found favor with those with whom he had discussed it. Rather, Gilbert wrote, "Newton stood with his two feet in two different currents of thought."[41] That image, of a Newton poised between two moments, would stay with Keynes even as he stowed the papers safely in a basement once the war began.

Christmas Day 1942 was the 300th anniversary of Newton's birth. Despite the increasingly desperate position of Britain in the war, plans were drawn up for a celebration of his life. Keynes loaned some of his manuscripts to an exhibition held at the Royal Society in London and was asked to give a talk.[42] He read his essay titled "Newton the Man" twice, once as an after-dinner address at the Royal Society on November 30, 1942, following a day-long appreciation of Newton (a bigger celebration was planned for and carried out after the war), and then on January 30, 1943, to the fellows of Trinity College, Cambridge, Newton's own college.[43] He had written his piece without the benefit of consulting his Newton manuscripts, which remained "buried in cellars" for the duration of the Blitz.[44] In correspondence he referred to the speech, drafted from memory, as "perfunctory and inaccurate."[45] But while Keynes did not include any references to specific manuscripts nor to specific events from Newton's lifetime, the piece is memorable. He was an almost irrepressibly spirited writer, and he knew, even without the primary materials in front of him, to what conclusions he had come and what would pique the interest of his audience, which included leading scientists such as the crystallographer William Lawrence Bragg and the physicist E. N. da C. Andrade.

His message was that Newton was not at all what he had been supposed to be. The Newton papers that he had rescued from an ignominious scattering contained secrets that Keynes was only too glad to reveal. The ideal Newton—the one who had emerged in the eighteenth century as "a rationalist, one who taught us to think on the lines of cold and untinctured reason"—was in for a sharp revision. The conclusion was unavoidable. "I do not think that any one who has pored over the contents of that box which he packed up when he finally left Cambridge in 1696 and which, though partly dispersed, have come down to us, can see him like that," proclaimed Keynes. Like Biot, and Baily before him, Keynes summoned up the private or secretive Newton, whose actions and beliefs were far from what might be expected of the idolized natural philosopher. Keynes's response to the theological and alchemical papers, the same ones that Liveing had catalogued back in the 1870s and 1880s, was not to dismiss them, as the Cambridge Syndicate had done, as befuddling and distracting, but to propose a complete revision in the way Newton was understood, to describe Newton "*as he was himself.*" It was not rational science that dominated Newton's worldview, argued Keynes, but a much older vision of the world as a union of connected parts. "Newton was not the first of the age of reason. He was the last of the magicians, the last of the Babylonians and Sumerians, the last great mind which looked out on the visible and intellectual world with the same eyes as those who began to build our intellectual inheritance rather less than 10,000 years ago."[46]

By "magic" Keynes meant Newton's putative belief in both natural and so-called mystical clues to the cosmic pattern of the universe. Keynes suggested that Newton "looked on the whole Universe and all that is in it as a *riddle*, as a secret which could be read by applying pure thought to certain evidence, certain mystic clues which God had lain about the world to allow a sort of philosopher's treasure hunt to the esoteric brotherhood." These clues were both natural (physical) and scribal, to be found in observation of the heavens and physical realities and partly "in certain papers and traditions handed down by the brethren in an unbroken chain back to the original cryptic revelation in Babylonia."[47]

On the old question of Newton's mental state, Keynes was certain: Newton's theological and alchemical beliefs were clearly the work of a sane man. The manuscripts were marked by "careful learning, accurate method, and extreme sobriety of statement." They were drafted, he concluded, during the twenty-five years that Newton lived in Trinity College—before the period of old age or any putative madness. Keynes seems to have been stumped, however, by how the same man could undertake such dramatically distinct forms of research and investigation.

Most of Newton's alchemical writings were notes and copies of other people's work, but even the chemical experiments to which he devoted hours could not redeem what was no more than an elaborate project to uncover the "riddle of tradition," not a serious scientific investigation. Based on the 100,000 words of alchemical translations and copies made by Newton that Keynes had seen, "it is utterly impossible to deny that it is wholly magical and wholly devoid of scientific value; and also impossible not to admit that Newton devoted years of work to it." The evidence for Newton's magical worldview provided an insight into Newton that Keynes concluded was ultimately (and seemingly paradoxically) uninteresting: impossible to discount, it was simultaneously impossible to credit. Ultimately Keynes concluded that the theological papers were of "no substantial value whatever except as a fascinating sidelight on the mind of our greatest genius." Having come so far in his analysis, Keynes was ultimately content to let Newton exist as a kind of freak who combined in one body and one mind the attributes of modernity and the premodern past. In this Keynes revealed the limits of his own imagination.

Though there was deviance in what drove Newton to seek patterns in both scriptural as well as natural forms, this deviance was also the key to his greatness, according to Keynes. He too commented on what Stokes and Adams had referred to as Newton's method—the means by which he achieved his indisputably great discoveries. Keynes's Newton was a solitary, hauntingly focused man whose powers of concentration enabled him to "hold a problem in his mind for hours and days and weeks until it surrendered to him its secret." This "peculiar gift" of thinking about something continuously was a talent that Keynes, the consummate polymath who flitted between ideas (albeit with the impressive grace and speed of a butterfly), could admire. Nevertheless Newton was, Keynes concluded, an extreme example of a "profoundly neurotic" individual whose appetite for the "occult, esoteric, semantic" led him to shrink unhealthily from the world. Keynes's own predilection for the trade in secret truths did not blind him to the dangers of such belief. Perhaps it made him more sensitive to them.

Keynes told the story of Newton's image and his papers together in one narrative for the first time. He dramatized Newton's papers in a way no one had. The Newton revealed by the manuscripts was a man of "secret heresies and scholastic superstitions it had been the study of a lifetime to conceal!" For his part, Keynes was willing, even eager to reveal what others had wished to keep secret. Newton's anti-Trinitarianism was the secret that his biographers had previously been most keen to keep. Keynes was matter-of-fact about revealing what he called Newton's "Judaic monotheism of the school of Maimonides." Keynes the book collector understood the significance of the history of those who had owned the papers before him, and he was not averse to placing himself publicly in that lineage. Disturbed by the "impiety" of the Sotheby's sale, he explained, he had managed to reassemble roughly half of the papers in a personal collection. Keynes had already arranged for his Newton papers to be donated to King's College in Cambridge, "which I hope they will never leave."

Keynes's time during the war was subsumed in advising the government. Suffering from ill health and overwork, he died in 1946, before he had a chance to revisit the Newton papers or revise his own essay. As he had intended, the books and papers he had collected were left to King's College, where they remain today. It is a measure of Keynes's fame and lasting influence that most of his writings have been published since his death, even the most quickly drafted and regrettable (such as a brief anti-Semitic sketch of Einstein).[48] That such embarrassing revelations were published says something about his editors' sense of fair play, perhaps, as well as their confidence in their subject's place in history. This also helps to explain the status accorded to the little essay that Keynes wrote on Newton. While the essay contained new insights and demonstrated Keynes's sensitivity to how significant the alchemical and theological work was to Newton, it was a biographical sketch, lacking detail and data. Nevertheless the essay has been influential in propagating an image of a "secret" Newton, more wild-eyed magician than clear-headed rationalist, as misleading in its way as the old myth of the scientist-saint.

12

# The Revealed Newton

Abraham Yahuda almost immediately grasped that the Newton papers that had been sold at Sotheby's were uniquely valuable, both as collector's items and as evidence of Newton's beliefs. Less than two weeks after the sale, he wrote to his wife, Ethel, that he was "now excited" about an unpublished essay by Newton on biblical and theological questions that had been sold at Sotheby's, which is "of the greatest significance for [Newton's] personal view" on matters of faith.[1] He set about trying to purchase the Newton papers and wrote to Ethel on July 28, "I am thrilled with the thought of acquiring them. He wrote a lot about the Bible and the Jews, about Cabbala and all sorts of Jewish questions."[2]

Nearly as quickly as Keynes had, Yahuda assembled his own Newton collection. He too secured good deals, from the dealer Gabriel Wells, who sold many lots to Yahuda at 15 percent commission, and from Maggs, who sold him a few lots at 20 percent commission. In his letters to Ethel, Yahuda boasts of "getting a great treasure which will be worth three times as much if not more very soon." At the time of the sale, Yahuda explained, the dealers did not realize how significant the manuscripts were, but that was changing fast. Yahuda, for one, seems to have immediately grasped their importance. "To have over 1500 pages written by Newton in his own hand on the most important questions is very thrilling indeed. But not only on Religion, Prophecies, Bibles, Faith, and Chronology, but also on alchemy, Mathematics and other purely scientific matters of the greatest importance for his studies and discoveries!!" It was almost too much of a good thing, and Yahuda, who hurried down to Victoria Station to give a check to Wells, would not "believe that I have the Mss. before I get them."[3]

Abraham Yahuda believed the theological manuscripts that he had purchased contained evidence of Newton's passionate drive to "extend the universalistic character of Christianity." Schwadron Portrait Collection, the National Library of Israel.

Yahuda was scathing about the "Museums & Libraries" that had failed to secure the manuscripts for Britain and bragged that with the upcoming tercentenary of Newton's birth in 1942, the papers would soon be worth five or ten times what he had paid for them. For all of August and September, as he set about acquiring as many lots as possible, Yahuda endeavored to keep the significance of what he was buying a secret, to keep prices low and the manuscripts coming to him. But it was already very clear to him that the papers revealed that Newton was "more a monotheist than a Trinitarian." In some parts of the manuscripts, Newton himself had concluded that "Jehovah is the unique god."[4]

Just weeks after the sale, Yahuda was forming an appreciation of what the papers contained that went beyond what almost anyone else had understood about Newton. Extremely quickly he was able to imaginatively integrate Newton's science with the new aspects of Newton revealed by the papers: his chronology and theology. He immediately rejected the notion that Newton's nonscientific writings were worthless. Like Plato's philosophy or Ptolemy's geography, he explained in a letter of August 30, "the 'results' are antiquated but the work bears the stamp of Newton's genius and it will always have value." The speed with which he came to this conclusion suggests that he must have been well-prepared to read the papers this way, but Yahuda himself felt that it was the papers that had changed him. "My occupation with Newton's papers have [*sic*] opened a new world to me and I am constantly under the spell of his personality," he confided to Ethel. "In these times of crises and ordeal he exercises a calming and reassuring influence upon me." As Yahuda was writing, the Nuremberg Laws had stripped Jews of their citizenship in Germany and Jews were forbidden from marrying non-Jews. In March Hitler had reoccupied the Rhineland, violating the Treaty of Versailles and raising the specter of war. The position of Jews in Europe was increasingly precarious. Yahuda grasped the redemptive potential of Newton's papers for the Jews, who could benefit from Newton's sympathy with their faith at a particularly vulnerable moment. More generally Newton's writings contained truths that could survive "destructions and Isolations." Yahuda found a way to hope through the papers: "Eternity belongs to the heroes of the spirit."[5]

Yahuda invested quite heavily in the purchase of the Newton papers. He spent more than £1,400 (more than £50,000 today) and sold some of his other manuscript stock to help fund the acquisitions. He didn't consider his acquisitions risky, though. What he had bought was of obvious value to him. The papers (he ended up, by his own estimation, with 3,400 folio pages) were the "best and most valuable" work he had ever purchased.[6]

Like Keynes, Yahuda had a certain claim to arrogance, to seeing truths that others could not. Where Keynes had been schooled at Eton and Cambridge, Yahuda's education contracted within its span Jerusalem, where he was born in 1877; Basel, where he attended the First Zionist Congress in 1897; and Germany, where he settled in for a series of degrees at Frankfurt, Heidelberg, and Strasbourg universities—all in the pursuit of knowledge suitable for the study of ancient texts and languages.[7]

The son of a rabbi, Yahuda was encouraged to study in the wider world but expected to observe Jewish law. An anecdote from his early education suggests something of his independence of spirit—and the tightly constrained world he inhabited, despite his travels. While studying in Frankfurt and living with an observant host family, he found himself unable to resist the urge to smoke a cigarette on the Sabbath. To soften the sinfulness of his act (and elude detection), he took a train to a nearby town. There he was unlucky enough to be seen by a relative. This almost comic breach had far-reaching effects: his observant family rejected him, and he thereafter embarked on a solitary life.[8] He was well suited for it. He published his first book, *Kadmoniyot ha-Aravim* (*The Arabs' Antiquities*) in 1893, when he was just fifteen. He continued his language studies in Heidelberg and Strasbourg, studying at the latter with the great Orientalist Theodor Nöldeke, who was clearly taken with the young scholar. A letter of recommendation written by his teacher describes Yahuda as a formidable linguist: "He not only speaks his native language, the Arabic of Jerusalem, but after becoming well versed in the written Arabic, he also acquired a thorough knowledge of the ancient and medieval Arabic language and literature." Nöldeke goes on to say that in addition to being a fluent speaker of German and writing it better than most Germans, his student had an "excellent" command of the Hebrew literature, was no "stranger" to Assyrian, and that it would be an "easy matter" for him to learn English.

By 1904, at the age of seventeen, Yahuda had his doctorate. The next year he put his prodigious language skills to use, teaching Semitic philology at a liberal rabbinical school and at the Orientalisches Seminar at Berlin University, where he stayed until 1914. He spent the next nine years in Berlin, eventually heading the Department of Biblical Studies and Semitic Languages at the university and lecturing on the exegesis of the Old Testament, a subject to which he would remain devoted all his life.

That illicit cigarette was not an anomaly; Yahuda's irreverent attitude persisted. He caused a stir by lecturing on the Bible without wearing a yarmulke. But his refusal to follow Jewish laws according to the letter did not mean he neglected the past. Instead he took a long view of history, seeking material to justify his own interpretations of tradition, however idiosyncratic.

In 1915 Yahuda was offered a professorship in rabbinic literature and languages at the University of Madrid, the first such position to be created in Spain since the expulsion of the Jews in 1492. Directed to make an appearance before King Alfonso XIII, Yahuda took the opportunity to proclaim both his heritage and his independence: "I am not the first in my family who appears in audience before one of your majesty's family," he informed the monarch. "It was in the mid-twelfth century, when one of my forefathers, Sheshet Benveniste, had the high honor of appearing before your majesty's forefather, King Alfonso II."[9] The appointment prompted newspaper articles proclaiming Yahuda's remarkable scholarly accomplishments, his common heritage with the Jews of Spain, and his tenacious devotion to his subject as guarantees of the wisdom of his appointment. He held his position for seven years, witnessing and participating in the extraordinary efforts of the international Zionist movement to secure a mandate in Palestine and making the first of a string of enemies in that movement, the beginning of a bitterness toward some of his fellow Jews and their ideas on how to run a Jewish state, which lasted his entire life.

Yahuda left Madrid in 1922 to embark on what would turn out to be a full twenty years of traveling, lecturing, and teaching. During this period he acquired a serious taste and facility for acquiring rare manuscripts, partially funded by money inherited by his wife, which led him to northern Africa, the Middle East, and western and eastern Europe. He taught in England at King's College, University College, London, Oxford, and Cambridge and lectured at such places as the Royal Asiatic Society of London, Hebrew University in Jerusalem, Yale University, and the University of Cairo. He spent the Nazi period in London, in the house on Ellsworthy Road where Keynes wrote to him. A visitor recalled extending his hand in the gloom of the foyer to "something formally elegant and stern," thinking it was the professor. Instead he encountered a life-size bust of the scholar, the man himself presiding in the reception room next door.[10]

Yahuda may have become a caricature of himself, but his learning was formidable. The fact remains that Yahuda, a Sephardic Jew born in Jerusalem to a family that had settled in Baghdad sometime after Spain expelled the Jews in 1492, was the first person to read Newton's private theological writings who was both able to understand them on their own terms and—perhaps more crucial—willing to do so. Back in 1777 Samuel Horsley may have looked at them, but he said nothing of it in his *Opera Omnia.* David Brewster and Francis Baily both saw them, and while they came to different conclusions about what the religious writings meant for understanding Newton, they both were concerned primarily with Newton the scientist. And whatever Luard, working in the 1870s on that long-delayed catalogue with Stokes, Adams, and

Liveing, may have thought privately about the writings, he went on the record to dismiss them as mere exercises in penmanship and evidence of an unhealthy obsession.

Yahuda turned the thing on its head. "His studies offer material about his concepts," he wrote in an unpublished essay on the Newton papers, "the manuscripts even more than the printed works." The theology manuscripts were not secondary; they offered a way of understanding Newton's scientific concepts that was ultimately more revealing than the printed works. What seemed "odd" in fact offered a "true estimate" of Newton. "It is necessary therefore that the remaining manuscripts are examined very carefully, so that the many things which appear odd today, receive their deserved vindication: this is a duty, not only towards Newton and his country but also towards all of humanity."[11] Yahuda was a scholar of history, pushing for the hard work of seeing the odd, old world that Newton inhabited.

Finally, all the excesses of the archive could be welcomed. It was the very extensiveness of the theological writings—the obsessive drafting, the lengthiness of the treatises—that spoke of Newton's desire to "extend the universalistic character of Christianity." By this Yahuda meant that Newton envisioned a truer, deeper religion, one that surpassed mere sectarianism, that "did not see the problem of religion exhausted in Christianity or Judea, but wanted to include all antique religions and the spiritual development of all other peoples besides the Israelites."[12] This was a message with obvious resonance in a Europe rent by war.

Rather than condescending to the eccentricities of an old man, Yahuda granted Newton this significant objective. It was the natural reason for the exuberance, not to say obsessiveness, of the writing. Rather than arguing against the drafts, as nearly all previous commentators had done, Yahuda could argue on their behalf. The drafts were evidence not of insanity or senility but of a vigorous faith that did not waver even in the presence of staggering amounts of historical data or scriptural language of the most obscure kind. The recopying, the fact that Newton had rewritten sections two or even three times, was evidence of the difficulty of the task as well as Newton's passion for it: "He strived for comprehension of the sense of the work and interpreted it completely several times with the highest degree of care."[13]

Yahuda's vision of Newton looks very different from any of the Newtons previously conjured by his would-be biographers, scholars, and cataloguers. But then Yahuda himself looked very different from any other Newton scholar. He saw language, and specifically scripture,

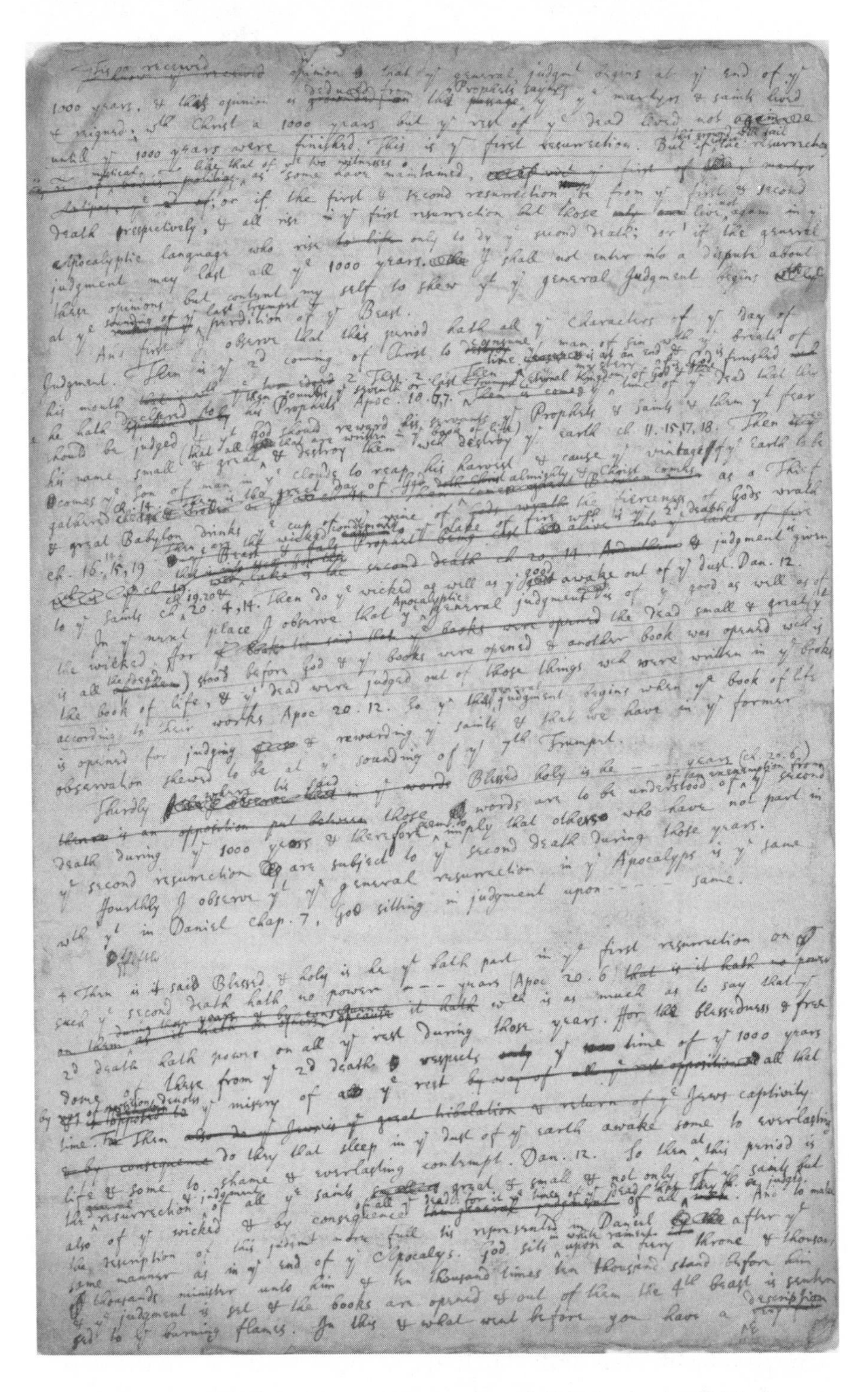

A page from one of the manuscripts purchased by Yahuda, in which Newton attempts to synchronize three parts of prophetic interpretation. Yahuda Ms. 6, f3v. Courtesy of the National Library of Israel.

as a code in which history itself could be read. Language bore the traces of lived experience, and great books, such as the Pentateuch, bore the traces of the history of the peoples it described. Like the tombs of the Egyptians, the Bible was a testament to history, and like the Egyptian hieroglyphs, the language within it could be studied to reveal its true meaning.

Yahuda was a forensic philologist, a practitioner of a brand of so-called Higher Criticism. "Lower Criticism" concerned itself with the nuts and bolts of transcription, the errors introduced into texts by lazy or unskilled scribes, the almost unavoidable mutations a manuscript underwent as it was copied over many years. Higher Criticism was after bigger game, capturing not simply the literal meaning of words written long ago but the entire worldview or culture in which those words were written. What did the writer of the text mean to accomplish at the time? What events surrounded its composition? Such questions seem plain enough, but when asked of the Bible, they become sensational.

Yahuda sought to trace the history of the Torah (the first five books of the Hebrew Bible) through telltale clues—words, customs, manners, and thoughts—borrowed by the Jews from the Egyptians, with whom they lived in close contact during their time in exile. He called this "proving the Hebrew-Egyptian relationship."[14] He was not the first nor the only scholar to seek to understand the Bible this way; what was distinctive about his approach was his emphasis on the Egyptian, as opposed to Babylonian or Assyrian, influences on the Bible.

In a way, it was precisely what Newton had sought to do: reduce the profusion of language in the Bible to a set of distinct meanings from which certain conclusions could be drawn. In the hundreds, if not thousands of pages on revelation and prophecy owned by Yahuda, Newton sought to translate the wild images and dense metaphors of prophecy into equivalent terms that were more stable. If he could fix the motile meanings of those writings into a set of equivalences, much could follow: confirmation of the contents of true religion, the truth of prophecy in the past, further evidence of God's omnipotence, and the true language of history. Similarly what Yahuda had done in his own work on the accuracy of the Bible was to demonstrate that the Jews had in fact walked over the ground it was said that they had, migrating from "Ur through Canaan to Egypt" and back to the Land of Promise. Incredibly enough it was possible to follow these long-effaced footsteps through the clues left in the language of the Bible itself. "In the development of the Hebrew language," Yahuda argued, "one can follow the route of Israel's wanderings during the last twenty-five centuries."[15]

For Yahuda, the Bible was a vibrant mix of cultures, practices, and perspectives whose very form testified to the cheek-by-jowl experiences of Jews, Egyptians, Babylonians, and Assyrians in the deep past. And yet he reserved a place of honor among the babble for the Chosen People who, he claimed, borrowed not so much "the vernaculars of primitive peoples in Israel's surroundings" but the "languages of the most cultivated peoples of the world."[16] This was not cultural relativism. For all his celebration of diversity, Yahuda was unafraid to assign prestige to the Jews, so long and so frequently abused.

This was not scholarship content to dabble in eccentricities and obliquities. To say that Yahuda was unafraid of controversy is not to say enough. He loved the brouhaha, the duels with other scholars. He lived a life at once public and solitary, giving many talks and participating in much correspondence but without much in the way of fellowship. According to his archivists, "his enormous correspondence contains few sustained relations."[17] Thanks perhaps to the legacy of his difficult personality, little has been written about Yahuda himself, though his archive of personal correspondence and manuscripts is extensive and his role in the intellectual universe of émigré Jews in the first half of the twentieth century substantial. And he seems to have grasped a truth about Newton that few had.

Abraham Yahuda's goal in acquiring manuscripts was ultimately historical: to reveal how biblical language had been shaped by the very history it described, or, as he put it, that "the Biblical narratives by their form, their style, their linguistic garb and peculiar colouring could only have developed in the course of the migrations of the Patriarchs from Ur through Canaan to Egypt and the return of the Hebrews from Egypt back to the Land of Promise."[18] He also wanted to claim the ground of Higher Criticism for himself, to take it back from practitioners such as Julius Wellhausen, perhaps the best known German biblical scholar of the time, who, in Yahuda's view, had taken things too far. "In the long run," he lamented, "it became customary to consider it as highly scientific to challenge everything Biblical and to alter the texts at one's heart's desire."[19]

For Yahuda this was a vision of criticism taken to extremes, the text reduced to nothing but error, the possibility of meaning dissolving amid a multiplicity of authors, leaving only commentary, a Talmud with no Torah left in it. He thought in particular that too many sources were being attributed to the Pentateuch and that too many "experts" were exerting themselves "in the art of text alterations and source-hunting."[20]

Thus "the original text was distorted and disfigured and in its place was offered a quite new text of pure invention."[21] In Newton, who himself sought to return a blemished Christianity to its purer origins, Yahuda found a kindred soul. Interpreting ancient texts didn't require robbing them of a fixed meaning. Both Newton and Yahuda sought instead to find a singular truth amid the variations. Even in the abstruse realm of textual criticism, much was at stake. As war raged in Europe, Yahuda made a case in a speech in New York for why it was so important to prove the accuracy of the Bible. Doing so was more than a "scientific concern"; it was a moral duty, so that the precious treasure that was the Bible was preserved from the "destructive theories of Higher Criticism" that may have contributed to the "spreading of those disruptive ideas which, to a large extent, paved the way in Germany to that 'Kultur' of racialism, paganism and self-deification as in the darkest ages of human history."[22]

Yahuda had legendary bouts of arrogance; he once half-seriously suggested that, on its founding, Israel should not be a democracy but a kingdom and he would be its first monarch.[23] But he had the admirable skill of assuming humility in relation to his texts. He made it clear that his writings on Newton were preliminary. Like all who encountered the papers firsthand, he understood the size of the task. "It is very probable that the picture and the purpose of Newton's religious studies will not become clear until all the material has been worked over and published," he wrote in an essay on the Newton papers. "There is no doubt, that, because only a part has been published, Newton's position has been misunderstood and the extent of the works has been underrated."[24] The manuscripts were more indicative, more revealing than anything Newton had published. Far from being irrelevant—or embarrassing or dangerous or a sign of lunacy—Newton's obsessive drafting was both the privilege and the curse of a man who was not free to publish in his own lifetime.

Though Yahuda's essays were never published, their intention is clear: to speak directly to those who rejected or ignored Newton's religious side and to counter the "scientific Newton" with the "outstandingly humanistic and moralistic figure." Yahuda's Newton is therefore almost totally distinct from the Newton of Brewster, Biot, Baily, De Morgan, or even Keynes, despite Keynes's interest in a "secret" Newton. Yahuda was unconcerned with distinguishing between Newton's supposedly good and bad selves, between his sanity and his faith, between his science and his heresy, between his science and his morals. More than an accumulation of simple dichotomies, Yahuda's Newton was based on valuations that were strange and wonderful. Rather than sections on his early life or the calculus dispute, Yahuda envisioned chapters in a proposed book that would cover Newton's "outer figure, face, voice, etc.," his "mood of life, his customs," and his knowledge of languages, of literature and fine art. His library deserved a chapter, thought Yahuda. So did Newton's handwriting, not least because it would aid in fixing a date to his manuscripts, the better to analyze the development of his "historical, chronological, religious" ideas. Finally, Yahuda suggested a section on Newton's "manner of working," of "making excerpts, of correcting and re-writing manuscripts."[25]

Yahuda's list raises the question of how to measure Newton: By the books on his shelves? By the habits of his note taking? By the features on his face? Taken individually, Yahuda's categories seem idiosyncratic, almost willfully so, but there is a serious intent to be read here, a vision of biography that hews closely to the practices of scholarship itself: to the books, the research, the handwriting, and the linguistic habits of its subject. Yahuda, a scholar who had no children but left a hefty estate of books and papers, was well-equipped to see the value of such a legacy. In his vision of how we might come to know Newton through his papers, Yahuda reveals a genuinely new way to look at the great man. If Yahuda's proposal seems almost aggressive in its exclusion of science, it has the merit of making Newton fresh and newly legible centuries after his death.

Yahuda's essays indicate that the value he attached to his Newton manuscripts was intellectual and cultural, not monetary. But a curious letter, written in 1940 as a testimonial by Albert Einstein at Yahuda's behest, provides a hint that he once considered selling them. Though it is unclear how they first met, Einstein and Yahuda had corresponded intensively from 1933 through the end of the decade, trading opinions on the increasingly dire situation in Europe and the political and diplomatic maneuverings surrounding partition plans for Palestine. Neither man mentions any of Yahuda's dealings in books and manuscripts, including his purchases in the summer and fall of 1936 of the Newton papers. Yahuda was more concerned with winning Einstein's support for his views on how matters in Israel should be conducted, while Einstein warned that "such polemics" would not be productive.

In early 1940 Einstein helped arrange for Yahuda and his wife to travel to New York. In the late summer of that year Yahuda visited Einstein at his summer retreat at Lake Saranac in the Adirondacks. The two men evidently discussed Newton, for preserved in the archives of both men is a letter in Einstein's hand dated September 1940 that details his views on Newton's private religious writings. Prompted no doubt by Yahuda, the document is notable for what we can infer of both Einstein's and Yahuda's attitudes toward the Newton papers. For while it is possible that Einstein made a sustained investigation of the papers that Yahuda owned, there is no evidence in either man's extensive personal archive that this was the case. Instead it seems much more likely that Yahuda was calling upon his famous acquaintance for a favor in helping him to dispose of his collection. A letter from Einstein declaiming the importance of the collection would serve as an excellent introduction to libraries that might be interested in purchasing the manuscripts.

"My dear Yahuda," wrote Einstein,

> Newton's writings on biblical subjects seem to me especially interesting because they provide deep insight into the characteristic intellectual features and working methods of this important man. The divine origin of the Bible is for Newton absolutely certain, a conviction that stands in curious contrast to the critical skepticism that characterizes his attitude toward the churches. From this confidence stems the firm conviction that the seemingly obscure parts of the Bible must contain important revelations, to illuminate which one need only decipher its symbolic language. Newton seeks this decipherment, or interpretation, by means of his sharp systematic thinking grounded on the careful use of all the sources at his disposal.
>
> While the formative development of Newton's lasting physics works must remain shrouded in darkness, because Newton apparently destroyed his preparatory works, we do have in this domain of his works on the Bible drafts and their repeated modification; these mostly unpublished writings therefore allow a highly interesting insight into the mental workshop of this unique thinker.[26]

Much as Stokes and Adams had before him, Einstein considered Newton's private papers with an eye toward gleaning as much as possible of his method of discovery, what he refers to here as "the formative development" of his work in physics. Einstein implicitly links the process by which Newton developed his physics and his theology; by studying the one, we might gain an insight into the other. He describes Newton's search for the secret truths of the Bible as deriving not from magical reasoning, as Keynes had determined, but from "sharp systematic thinking." Newton's so-called mental workshop (*geistige Werkstatt*) is metaphorically the same place where both his physics and his theology were created. Drafts are in no way evidence of dangerous obsession or weak-minded repetition but the evidence of a mind at work on the way to creation. Confirming the likelihood that Yahuda was seeking a testimonial to help him sell the papers to a library, Einstein added at the end of his letter that he considered it "highly desirable, that Newton's writings mentioned above be united in one location and there be made accessible to researchers."

While Einstein was happy to help Yahuda try to sell the papers and to support the cause of scholarship, later in his life he expressed a different view on the proper disposition of the papers. In an interview with the historian of science I. B. Cohen that took place just two weeks before he died in 1955, Einstein spoke about, among other things, Newton's theological writings. He said that it was significant that Newton had "sealed them all up in a box," an indication, Einstein thought, of Newton's awareness of how imperfect they were. Newton had "obviously" not wanted to publish these speculations during his own lifetime. Einstein said, "with some passion," that he hoped they would not now be published. Speaking as someone who had lived most of his life squarely in the public eye, he defended Newton's right to privacy, even after death. Rather than lamenting the absence of a complete edition of Newton's works, Einstein praised the Royal Society's resistance to publishing writings that had remained unpublished during Newton's life. Correspondence could reasonably be printed, having been made somewhat public during Newton's life, but there was always the possibility that the letters contained certain "personal things which should not be published."[27]

Despite Einstein's letter, Yahuda never did sell the Newton papers. In 1942 he traveled as a refugee to America, where, like so many other scholars uprooted by the war, he found a place at the New School for Social Research in New York. For several years he ran the school's Center for the Study of the Near and Middle East and gave lectures on biblical literature, Islamic architecture and ornamental art, Semitic inscriptions, advanced Arabic, and a survey course on the history of the ancient Near East. No evidence suggests that he shared his Newton papers with his students.

Yahuda and his wife moved in the last years of his life to New Haven, Connecticut, and though he was not on the faculty at Yale University, he hoped to make connections with scholars there. Instead he was to be "lonely as never before."[28] In addition to the Newton papers, Yahuda had acquired over the course of more than forty years of collecting what was reputed to be the largest and most valuable assemblage of rare Arabic books and manuscripts in private hands. The bulk of these—an incredible 4,800 Arabic texts spanning a thousand years of history and ranging across astronomy, mathematics, literature, geography, philosophy, and medicine—ended up in the Princeton University Library, making it the largest repository of Islamic manuscripts in North America (which it remains today). Yahuda also sold some Arabic medical manuscripts to the U.S. Armed Forces Medical Library in Washington, D.C., as well as additional materials to Dublin's Chester Beatty Collection.[29]

In August 1951, while on vacation with his wife at Saratoga Springs, New York, Yahuda died of a heart attack at age seventy-four. His obituary appeared the next day in the *New York Times*, hailing him as a "noted expert on the Bible and orientalist." He had died in many ways an isolated and angry man. His "learning was immense," as the *Times* obituarist noted; his "reasoning and judgment, however, were not consistently sound."[30] The dispersal of his collection after his death was to be as fraught as his relationships were in life. Before he died he had sent some books and papers from his library to a warehouse in New Haven to await packing for shipment overseas. But he never packed them up or designated a recipient for them. Instead the books sat in the warehouse for several years until Ethel Yahuda, who on her husband's death had inherited the entire library, valued at $80,000, began to prepare it for shipment. Despite Yahuda's lifelong anti-Zionism, a result of his deep discord with the noted Zionist and Israel's first president Chaim Weizmann, among others, Ethel decided to donate all of Yahuda's books and manuscripts—including the Newton materials—to the Jewish National and University Library at Hebrew University in Jerusalem. She had been convinced by a Boston book dealer named Abraham Bornstein to honor the people of Israel with a bequest. More important than any quarrels Yahuda may have had during his lifetime was the legacy he could leave, in the form of his books and manuscripts, to the Jewish people.[31] She made the announcement at a luncheon in Israel (which was attended by the president of Israel) on January 28, 1953. Soon afterward she began to arrange the material that had been sitting in the warehouse for so many years; the cataloguing and crating of the material was still unfinished at the time of her death in 1955.

Though she had publicly announced her intention to make the gift, Ethel had made no written provision in her will regarding a donation to the university. One of the four trustees of the estate, a nephew of her late husband, objected to the donation of the library to Hebrew University. The resulting court case dragged on until 1966, when the Supreme Court of Connecticut ruled that the library should be donated to the Jewish National and University Library after all, having found that Ethel Yahuda's intentions had been clearly stated orally to a number of people before her death. The case of *Hebrew University Assn. v. Nye* has subsequently served as an important precedent for honoring the intention of a donation in the absence of written documentation.[32] The fate of the collection of Abraham Yahuda, a man for whom written language carried the promise of revealing deep and lasting truths, depended ultimately on spoken words. Following the court's decision, the collection, including all of the Newton papers, was finally crated and shipped to Israel.

13

# The Newton Industry

Newton's papers were now out in the world. Dispersed at the Sotheby's sale in 1936, the majority of the nonscientific material ended up back in institutional hands—principally in Cambridge, in Jerusalem at the Jewish National and University Library, and in Kew at the National Archives—thanks to the combined efforts of Keynes and Yahuda (and the gentlemanly actions of many of the dealers). But though they were available for the first time since Newton wrote them, both the nonscientific and the scientific papers were to remain almost entirely unexamined by scholars for nearly two more decades.

The question is why. In 1924 J. L. E. Dreyer, president of the Royal Astronomical Society, delivered a speech in which he addressed the matter of Newton and the lack of proper editions.[1] He remarked that one had to go back to Samuel Horsley's *Opera Omnia* (1779–85) to find anything "approaching completeness," but he quickly added that Horsley's edition was lacking even in its day and could not come close to fulfilling modern editorial and scholarly requirements. Other nations had managed to produce fine editions of a long list of scientific heroes, including Copernicus, Tycho Brahe, Kepler, Galileo, Torricelli, Descartes, Fermat, Huygens, Leibniz, Euler, Lagrange, and Laplace. In 1912 Dreyer himself had edited a collection of the scientific papers of William Herschel, the German-born British astronomer who had discovered Uranus in the late eighteenth century. The absence of Isaac Newton from the list was conspicuous and damning.

Dreyer called attention to the fact that the entire collection of scientific papers that had been catalogued by the Cambridge Syndicate was lying unremarked in the University Library, along with correspondence in Trinity College, in the private collection of Lord Macclesfield, at Corpus Christi, Oxford, and elsewhere. (He mentioned only in passing the theological and alchemical material that was then still in Hurstbourne Park.) Surely, he implored, it would not be so difficult to gather together the relevant material and create an edition that would give due appreciation to the proudest flower of English science.

Dreyer's confidence met with a sobering response in the very same edition of the Society's journal. R. A. Sampson, the erstwhile editor in chief of a misbegotten (and long forgotten) attempt to do honor to Newton back in the early years of the century had his memory, and his conscience, piqued by the president's speech. That previous project, undertaken in 1904 by Cambridge University Press with Sampson as editor in chief, had been rather breezily estimated to require just six years to produce six volumes of about six hundred pages each. It was planned to include all the manuscripts one might reasonably hope for, including correspondence held at the Royal Society, the Portsmouth papers (presumably those at Cambridge and excluding the nonscientific papers at Hurstbourne), miscellaneous other letters held at libraries throughout England, and an edition of the *Principia* in English (not printed since A. Motte's 1729 edition). These plans were soon stymied by other work commitments and a growing recognition that such an edition was far more challenging than anyone had realized. The undertaking foundered before it got off the ground.

Looking back twenty years later, Sampson was much less sanguine than he had been in 1904 about the feasibility of such an edition. There were problems on all sides. He identified the paradox of all great scholarly editions: useful to a few historians, they were useless to the vast majority of people. "It is an historical task, and, in a sense, one of national importance, but for the advancement of science it matters only in moderate degree." The costs to the nation might be deemed worthwhile; the potential personal costs to the editor were not to be underestimated, for editing such a work had many risks, some of them existential. "Before anyone undertakes it he should see clearly that he is making a voyage into the past, the greater part of it completely dead and obsolete," wrote Sampson ominously. "An editor must allow himself to be absorbed into this defunct world for many years. Probably he would never quite find his way out of it." Certainly the right man for the job needed scientific understanding, but he also crucially needed the ability to resist the potentially dangerous features of the manuscripts. The Newton archive was, in Sampson's description, almost as impregnable as Sleeping Beauty's castle: "Examined under a microscope, Newton's work shows no cracks or crevices. In consequence, everyone who has worked on Newton has fallen under a spell."[2]

Though Sampson's view now seems stuffy—he had a hard time taking the alchemy seriously as a subject worthy of either the historian or Newton himself—he understood something significant about the archive: the deeper one went down into it, the harder it was to hold on to old certainties about what knowledge contained there was really valuable. It was as much a maze as a fortress. The closer one looked, the more it all seemed interrelated. This was a dangerous insight, for it linked natural with magical laws, and Sampson ultimately drew back from the precipice. While the interconnections were real, they were also, he concluded, of no use to anyone—not the editor, not the scholar, not the reader of what might be written on them. What was valuable instead was finding the origins and the development of the great scientific insights. Everything else, no matter how linked and how intricately interwoven, had to be discarded: alchemy, theology, administration, all of it. The challenge for the editor, as Sampson saw it, was to put each aspect of the archive into its proper relation to all the other pieces, to exhibit "judgment as much as knowledge" in sorting out the wheat from the chaff. Faced with a likely fate of years lost in the maze of Newton's seamless archive, Sampson concluded that only a professional historian should undertake the task, and that historian would have to be interested in such statements as "a ray of light has four sides." The problem quickly became insoluble. The best that Sampson could advise was that a "senior man" be found to take on the task, someone who had both more authority and less in the way of his future career to lose from such a godforsaken project. Sampson's heartfelt and conscience-burned essay is all the more poignant for his failure to even mention the discipline of history of science. He was not to blame. In 1924 the number of professional historians of science at work in the Anglophone world was countable on one hand.

As Dreyer mentioned, other nations had managed to publish monuments to their great scientists, despite the editorial challenges. Starting in the final decades of the nineteenth century, the Latin phrase *opera omnia,* meaning "complete works," had appeared with increasing frequency next to the names of great Continental natural philosophers from the same historical period as Newton. In some cases these national heroes were to receive the editorial equivalent of a knighthood two or even three times over. Italy, for example, had honored Galileo many times, first with a fifteen-volume edition (1842–56) and then a twenty-volume edition (1890–1909, reprinted with additions in 1929–39 and 1964–66). France had honored Descartes in a famous thirteen-volume edition (1897–1913) by Charles Adam and Paul Tannery, and Fermat had been the subject of another five-volume edition (1891–1912), also by Paul Tannery with C. Henry.

Other projects had been commenced and remained ongoing. The Dutch had begun a project on Christiaan Huygens in 1888 that would be completed (in 1950) in twenty-two volumes after more than sixty years of labor had been devoted to the task. The Swiss had inaugurated their homage to Leonhard Euler in 1911 with a project that is still under way (eighty of a projected eighty-four volumes have been finished to date, more than a century after the start). Perhaps the most impressive of these projects was undertaken in Germany, which sponsored two complete editions of Kepler, one in the mid-nineteenth century (eight volumes edited by Christian Frisch from 1858 to 1871) and one in the twentieth (twenty-five of a projected twenty-six volumes completed, begun 1937 and still continuing to date). Last, the truly monumental 120-volume project to edit the papers of Gottfried Leibniz, begun in 1923, is also still under way.

While these Continental projects proceeded, the papers of Isaac Newton waited, like a fairytale princess waiting for the prince who would bestow the kind of kiss to end centuries of enchantment. Why Britain hadn't produced anything of note to honor its greatest scientist was a question that had been sporadically asked, to little avail. One reason was that in Britain, the government preferred to let individuals drive the scientific research agenda and traditionally it had been unwilling to fund all but the most exceptional large-scale scientific projects, much less those relating to the obscure field of history of science. In contrast to the Continent, where a long-term undertaking to produce a critical edition of a famed scientist could be considered an effort in state building, in England it was considered a wasteful application of state funds. In the absence of government commitment to produce a critical edition of Newton for the nation, what the papers really awaited was the birth of a new professional discipline.

One man in particular was responsible for bringing the history of science out of the imaginary and into the real world. In 1914 George Sarton was a young Belgian scholar in chemistry and mathematics who had just begun to take notes for an ambitious new historical approach. In advance of the arrival of the German Army, he buried these notes in the front garden of his house in Brussels. Sarton survived the war, and so did his notes, which he used to write an opening manifesto on the need to found a new field of human inquiry. For Sarton, the history of science fulfilled an almost messianic role as a form of "universal" history that had the potential to knit mankind (back) together following the barbarity of war.

Sarton's proposal was systematic and bristling with self-confidence, even in the face of war and the staggering amount of work to be done. He understood, as had Sampson, both the importance of the task at hand and the utter obscurity of the endeavor. The wider world may have had no idea what was at stake, but the importance of the matter remained undiminished. What lay in the balance was nothing less than an understanding of civilization itself. At a time when human culture seemed poised on the brink of destruction, the history of science offered an essential form of civilizational self-knowledge. To Sarton it was clear that "the history of science is even more significant than the history of religion or the history of art, which have long enjoyed full recognition as separate branches of learning and have reached a high degree of organization." Unless historical and scientific information were combined, knowledge of nature and of man would not be complete. The history of science was nothing less than "the keystone of the whole structure." Sarton's conviction did not blind him to the fact that he was largely alone in his endeavor. He admitted that the number of scientists and historians who shared his belief was "exceedingly small."[3] Undaunted, he devoted his long life to erecting the scaffold of such a discipline.

After the war Sarton moved to the United States and began his project with evangelical zeal. He founded a scholarly journal (*Isis*), a scholarly society (the History of Science Society, the long-awaited inheritor of the mantle shed by Halliwell's abortive Historical Society of Science back in 1846), and, somewhat more tenuously, a scholarly department (at Harvard, where he waited many years before eventually being granted full tenure in 1940). Sarton had a martyr's fervor in the face of impossible odds. He wished "to be able to carry the load many more years, and in any case I hope that I may be privileged to die before I break down."[4] Believing as he did in the need for "total history," Sarton began at the beginning. He aimed to write nothing less than the entire history of science. He was to reach the fourteenth century before he did finally break down.

Back in Europe another man was working his way toward Newton, traveling not chronological but philosophical avenues to reach his man. The Russian-born Alexandre Koyré, a student of Edmund Husserl and product of prewar Continental philosophy, came to see the "Newtonian synthesis" as one of the "deepest if not the deepest mutations and transformations accomplished—or suffered—by the human mind since the invention of the cosmos by the Greeks, two thousand years before."[5] For Koyré, it was theory, not experiment, that drove discovery. Together Sarton and Koyré helped to define what the history of science would concern itself with. Yet while they helped bring the discipline of history of science into being, neither betrayed any interest in the archival basis for history. Sarton had been preoccupied with forging the institutional foundation for a sweeping vision of history. Koyré set out to extract from the fleshy mess of history the sinews of an intellectual history. He outlined revolutions in thought, tracing a genealogy that stretched from Copernicus to Kepler to Galileo and culminated in Newton.

While Sarton and Koyré had helped to create the conditions under which scholars might think more deeply about Newton's role in history, they left it to the next generation of scholars to push beyond published texts and into the unpublished manuscripts. Newton was a natural subject for a discipline in search of itself. He was, after all, the crowning glory of the Scientific Revolution, the seedbed for the discipline. Nothing could be more important than understanding, via the Scientific Revolution and primarily via Newton, the origins of modern science and its contribution to the origins of modernity itself. And so, beginning in the 1950s, the Newton papers finally began to receive attention from a series of scholars who saw in the papers both the means for professional advancement and the answer to certain historical and intellectual questions recently posed that seemed increasingly urgent. Though famously uninterested in students, Sarton had managed to cultivate one who would take on his role as the patriarch of the discipline in America. In 1947 I. B. Cohen (the "I" stood for precisely nothing, he claimed, though he admitted to "Bernard") was the first American to receive a doctorate in the history of science.[6] Succeeding Sarton as both editor of *Isis* and president of the History of Science Society, Cohen lived a long and fruitful life, during which he witnessed and helped to bring about the transformation of the new discipline from a weedy and highly specialized subject to one that seemed to offer a way to knit back the two cultures—the sciences and the humanities—which, C. P. Snow had famously lamented, had become dangerously divided.[7]

Sarton himself had not written anything on Newton save for a few reviews. He was too busy trying to write the entire history of science.

I. B. Cohen believed that it was important to reproduce "the pre-drafts, the early versions, the stages of successive alteration" when publishing the Newton papers. With Alexandre Koyré and Anne Whitman, he produced a variorum edition of the *Principia* in which as many changes as possible were included. Courtesy of the Schlesinger Library, Radcliffe Institute, Harvard University.

Cohen got to Newton comparatively much faster. He began his dissertation in 1937, and just five years later published an essay entitled "Newton and the Modern World" in the *American Scholar*. His first monograph, published in 1956, compared Newton and Benjamin Franklin and showed how Newtonianism had come to America. It was then that his study of Newton began in earnest. Over the next three decades Cohen would be a driving force in a growing group of Newton scholars, even as he helped to shape the broader discipline of the history of science and promote it to the world. During the 1950s and 1960s scholars produced a bumper crop of editions containing never-before-published manuscript material. These included Herbert McLachlan's *Theological Manuscripts* (1950), which contained documents from the Keynes collection but nothing from the much more extensive Yahuda holdings in Jerusalem; the *Unpublished Scientific Papers of Isaac Newton* (1962), edited by the husband-and-wife team of A. Rupert Hall and Marie Boas Hall, which included a selection, far from comprehensive, of the Portsmouth papers that had ended up in the Cambridge University Library; J. W. Herivel's *The Background to Newton's Principia* (1965), which made public a number of primary source documents relating to work that preceded the *Principia*; and two major editorial enterprises. The first of these, which appeared in 1959 under the auspices of the Royal Society (making good on the abortive attempts to edit the letters in 1904 and 1924), was *The Correspondence of Isaac Newton*, which had many editors but was finally finished under the supervision of the Halls; the second was D. T. Whiteside's heroic, nearly single-handed eight-volume edition of the mathematical papers, *The Mathematical Papers of Isaac Newton*, the first volume of which appeared in 1967.[8]

The Newton papers were, finally, well and truly out of the dark and in the hands of scholars with the enthusiasm, skill, and professional motivation to begin to make sense of them. But despite the energy that went into cataloguing them, the papers were still formidable, even exhausting. The weariness of would-be cataloguers and editors of the papers echoes down through the ages. In 1959 the first volume of *The Correspondence of Isaac Newton* (covering the "genius" years of 1661–75) appeared, and it was with an almost audible sigh that editor E. N. da C. Andrade estimated that among the "mass of unpublished documents" that Newton had left behind were some 1.4 million words on theology and biblical chronology, 550,000 on alchemy and related subjects, about one million on the sciences, 150,000 on coinage and the Mint, and at least half a million on "subjects difficult to classify," yielding a grand total of 3.6 million words. Andrade ventured that to do them justice would take up twenty-five volumes. Much of it, including Newton's copying of drafts and lengthy calculations, noted Andrade, was excessive, tedious, and tiresome: "A large, possibly very large, proportion of this manuscript matter is, then, of little value, and would certainly be rejected altogether by wise editors; but it is hard to suppose that there are not treasures among it, and to separate them and to identify the copyings would require long labour of many patient and profound experts such as it is hard to find today."[9] Andrade's "but," though honest, was like a baited fish hook to an ambitious scholar. Such was the peril of the scholar, dismayed by his task but reluctant to discard anything lurking within that might be valuable.

Overall Andrade's tone was discouraging. Much of Newton's mathematical work had been published in a "delayed, scattered, and in some ways imperfect manner." A critical edition of the *Commercium Epistolicum* would be "a very troublesome task." The *Principia* alone would make "very heavy demands upon editors." Moreover the age of leisurely scholarship in which a great edition (such as those produced on the Continent) might have been accomplished, Andrade suggested, had already passed.[10]

As the appetite for primary source material grew, questions arose about how much was too much. How much did editors have to convey about the materiality of the manuscripts? Did every addition and deletion need to be shared? How many versions of the same text should be published? Underlying all these was the question of what manuscripts were good for. Cohen argued that, in the case of Newtonianism in the eighteenth century at least, print rather than manuscripts was what mattered.[11] But increasingly the line between print and manuscript was seen as a ragged edge. Unintuitive though it seems, in the seventeenth and eighteenth centuries, manuscripts often circulated widely, while printed works, such as the *Principia*, might have a severely limited audience. Newton had given at least a dozen people access to his unpublished papers during his lifetime. Manuscripts were loaned, copied, and recirculated, sometimes with and often without Newton's permission, increasing readership in ways that were significant if hard to quantify.

Better equipped in every way and attempting to achieve just a fraction of what early Newton editors had envisioned, Cohen nevertheless experienced the same sense of vertigo that Sampson and others before him had described. In his *Introduction to the "Principia,"* Cohen took considerable time to explore the editorial problematic. He explained how, following World War II, the goals of the history of science had dramatically changed. No longer was it sufficient merely to enumerate great discoveries, as Koyré had primarily done. Historians were henceforth concerned with what Cohen called the "growth of ideas rather than their mere final expression." The importance of primary documents—books, manuscripts, and even scientific instruments—was concomitant and necessary. The vision of the history of science that Cohen laid out was of a breathless, restless, and constant delving into the past, an incremental exhumation of "the pre-drafts, the early versions, the stages of successive alteration."[12]

As an example, Cohen explained how successive biographers had treated as a single unit the "Queries," a series of philosophical questions on the nature of light, heat, electricity, gravity, and chemical affinity, among other things, that appeared at the end of Newton's *Opticks*. But, as Cohen pointed out, the Queries had evolved over more than a dozen years. There is no single set of "Queries" as such; depending on which edition you look at, there are sixteen, twenty-three, or thirty-one. Having come to the understanding that it is only by studying the "gradual unfolding" of Newton's thought that the true nature of his mind could be understood, Cohen encountered a conundrum of potentially infinite extent—that of how many "stages of successive alteration" could feasibly be included in any edited work. Where should one draw the line?

With the assistance of Latinist Anne Whitman and in collaboration with Koyré, Cohen set out to create a so-called variorum edition of the *Principia*, including all three printed editions (dating from 1687, 1713, and 1726), as well as the final draft of the first edition submitted to the printer and all annotations made by Newton to the three editions.[13] The work traced every change that could be linked to a definite proposition in the *Principia*, showing in unprecedented detail how extensively Newton had revised his work. However, such a project, ambitious as it was, still captured only so many discrete moments in time. Great though the mass of annotation assembled in the variorum edition, there was greater subtlety still in the evolution of Newton's thought. Growing evidence from annotated and interleaved copies of the editions deepened the understanding of Newton's science in significant ways. But it was simply not possible to do it all, however desirable. The "sheer bulk" of the changes made it painfully apparent that "the task of preparing an edition with absolute completeness was beyond our capacities as editors." Among other things, the project would always be haunted by all the material that had been lost or destroyed without leaving a trace. Sounding more and more like Sampson, Cohen argued that it would take a dozen scholars a dozen years to carry out a truly comprehensive edition, which would be "an unwise expenditure of our presently available scholarly resources of both manpower and funds." Even should a thing prove possible to create, it would be massively expensive, overly detailed, and of use to none "but the most dedicated specialist."[14]

Cohen had to admit—almost despite himself—that there were limits to what could be known. Despite the reams of material both published and unpublished, "there are many interesting and important questions that cannot be answered fully or definitively, and some others cannot be answered at all."[15] What is extraordinary about this statement is that Cohen felt bound to make it at all. The implicit assumption is that the Newton papers might somehow answer any question we might put to them. The explosion of new material available on Newton, some 250 years after his death, had made it seem that not only were there no unanswerable questions but that the answers to questions as yet unasked lay there, waiting to be discovered.

The survival of the material was remarkable. In the early days of the Royal Society John Aubrey had denounced his fellow men of science, antiquaries, and philosophers for allowing their papers to be turned into pie-wrappings and fire-starters. Some sizable collections of papers had survived, thanks perhaps to the exhortations of Aubrey and others, and these included the manuscripts of such important figures as Galileo, Robert Boyle, Huygens, Leibniz, Flamsteed, Locke, and Samuel Hartlib.[16] Yet much of what was created by Newton's peers was also lost to time. The papers of the renowned physician William Harvey were lost during the political upheaval of the 1640s, while those of Descartes survived shipwreck after his death only to be lost in the eighteenth century. Newton took care to preserve the majority of his papers during his life, and his illustrious position as master of the Mint had helped raise his half-niece's social status to the point where she could marry a man as well-born as John Conduitt, thus helping to ensure a noble line of descent for the paper progeny of the childless philosopher. Nonetheless the survival of the papers was a happy example of the winds of archival fate blowing in the right direction for nearly three hundred years.

By the mid-1970s the history of science had established itself as a new discipline full of energy and enthusiasm for the material as well as intellectual products of history. Within it the special subset of Newton studies had become as robust and prolific as a factory production line. Despite the maturation of the field, the special community of Newton scholars that sheltered within it remained eccentric. Most of the men (and they were nearly all men) who devoted time to Newton's manuscript legacy—scientific as well as nonscientific—were, in some sense, outsiders. Cohen, as mentioned, was the first American to earn a doctorate in the history of science. D. T. Whiteside had no advanced mathematical training when he undertook the editing of Newton's mathematical papers and worked largely alone to produce his eight volumes. Keynes had no obvious reason to be interested in Newton, much less his alchemy. And Yahuda had alienated himself from every group he might ever have been accepted into. But even against this landscape of outliers, David Castillejo, who would be instrumental in bringing to light a large portion of Newton's manuscript legacy, stands out.

Born in Spain in 1927, the third of four children, to a Spanish father and an English mother, both active in educational and political reform, Castillejo moved with his siblings to London during the Spanish Civil War. By 1947 he was at King's College, Cambridge, studying seventeenth-century metaphysical poetry. His older brother, Leonardo, who went on to become a well-regarded physicist, also at Cambridge, mentioned that Keynes's papers had been recently bequeathed to the college. At that time A. N. L. Munby, the same man who as a cataloguer for Sotheby's had described the experience of assessing country house libraries for sale and who had bought books from Gustave David in the Cambridge market square, was the college librarian. Castillejo remembers them together spreading out on the big library tables the Newton materials that Keynes had collected. He followed up on the investigation by traveling to Paris and speaking with the dealer Emmanuel Fabius, enjoying an excellent meal while Fabius told his story of attending the Sotheby sale.[17]

Back in Cambridge, Castillejo pored over the materials Keynes had gathered. He was almost certainly the first to do so since Keynes, and one of very few ever to have studied the papers at all. He realized that the papers were notes on alchemical readings, notes that seemed to be taken with the aim of preparing a theoretical text on alchemy. Castillejo guessed there was more to Newton's alchemical project than was contained in these papers. Where, he wondered, was the record of the alchemical experiments that must have accompanied Newton's paper research? Nothing of the sort was to be found among Keynes's bequest.

On a hunch, Castillejo went to Cambridge's University Library and made inquiries about the material they had received in 1888. Was there anything resembling notebooks of alchemical experiments? The librarian looked around and found only a small batch of papers. However, he mentioned that a notebook had been recorded when the papers were originally received but seemed to have been misplaced during a move from the old library building to the new University Library site. Following the trail, Castillejo, with Munby's help, tracked down a Mr. Baker at the Old School's library who managed a crucial, and prodigious, feat of memory.

As Castillejo himself recalled more than sixty years after the event, Baker reported that he did remember having seen in 1914 a "brown paper parcel" with the word "Newton" on it. The difficulty was that the library had been "re-arranged three times" since then. Nevertheless Baker, good librarian that he was, tried once, twice, and, in 1949, a third time to locate the parcel. Castillejo was in the final stages of preparing a fellowship thesis for the college at this point. On being asked by Castillejo if he should report the notebooks lost, Baker answered that given the next day was Saturday he would have another look. Wonder of wonders, he came up with the goods. Castillejo sat down with the notebooks to study "for as many days as I liked in the great empty red room of the Senate next to the library." Today those notebooks reside in the University Library alongside another, similar notebook of experiments that was found with it. They reveal just how much time Newton spent in his chemical laboratory over the course of thirty years. Scholars have used them to re-create some of Newton's actual experiments to better understand the nature of his alchemical investigations and of early modern chemistry and alchemy more generally.[18]

In the notebooks, Newton recorded more than fifty thousand words' worth of notes on alchemical experiments. The experiments are devoted to the alchemical pursuit of wresting meaning from complex imagery and the complicated interactions of minerals and chemicals. The goals of this activity were many: for some it was the old hope of transmuting base metals into gold, but for Newton it was understanding an aspect of the world—its secret, vital forces—that remained hidden to all but a few. Such knowledge was not easily gained, and Newton devoted countless hours to it. He returned repeatedly to several key alchemical puzzles. One of these was the star regulus of antimony, a metal made by slowly cooling melted antimony until it creates a visible crystalline structure, the so-called star. Newton tried it as a reflector in his reflecting telescope, and though he soon rejected it for that purpose, he remained faithful to it for much longer in his alchemical work. Another alchemical mystery was a purple alloy of metallic antimony and copper. Following the recipes of Eirenaeus Philalethes (the pseudonym of the American alchemist George Starkey), Newton sought ever more successful methods of refining this alloy, which was named the "net" after the tiny regular crystals on its surface that resembled a network.[19]

Successful in his hunt for the laboratory notebook, Castillejo finished his thesis but failed to get a fellowship that might have allowed him to stay at Cambridge. He subsequently left the university. His engagement with the forgotten manuscripts of Newton did not end there, however. Having been alerted by Munby that there were Newton papers in Jerusalem, Castillejo found occasion to visit in 1969. He

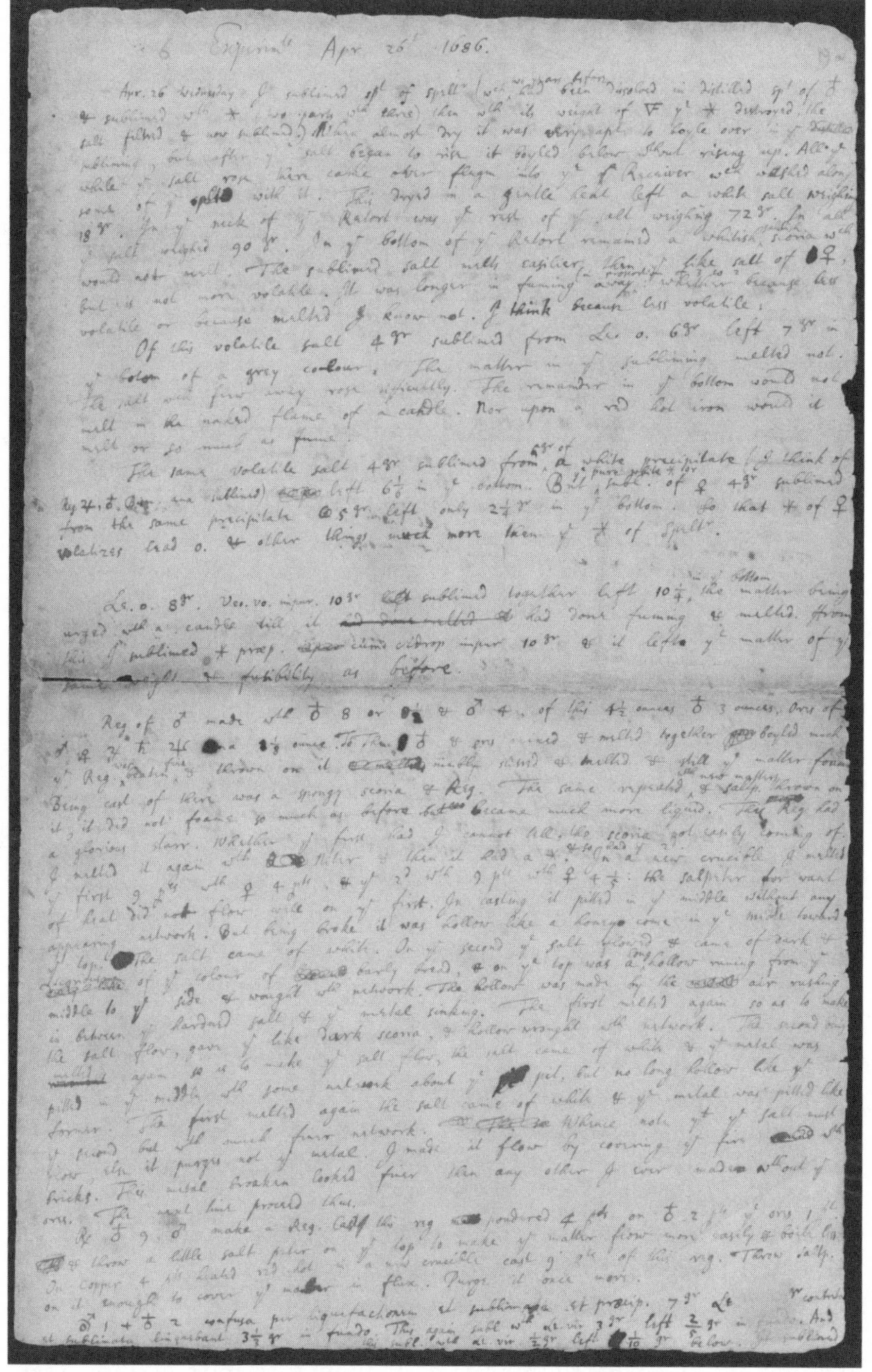

A sheet of notes from one of Newton's alchemical notebooks in which he documents his attempts to create the star regulus, a crystalline structure obtained by slowly cooling the metal. Misplaced for nearly seventy years, the notebook was discovered in Cambridge in the 1940s by David Castillejo. MS Add. 3973 f19r. Reproduced by kind permission of the Syndics of Cambridge University Library.

simply turned up and asked to see the librarian, Mordekai Nadav, leafing through the card catalogue as he waited. The librarian was shocked when Castillejo asked him about the Newton papers, since they had only just barely arrived and were hardly unpacked, much less catalogued. Castillejo demonstrated his knowledge by inquiring after a specific document according to its number in the Sotheby sales catalogue, a classification system that he and Munby had relied upon when organizing the Keynes papers. Impressed, Nadav invited Castillejo back on Monday. When he arrived, Nadav greeted him with a trolley packed with bundles of Newton's manuscripts. "Would you catalogue them for me?" he asked. "We know nothing about Newton here." So began Castillejo's second adventure with Newton's private thoughts. He spent two weeks in Israel cataloguing the papers and many years thereafter studying what he had uncovered.

While Castillejo worked alone in Jerusalem, in Britain there was an upswell of interest in all things Newton. D. T. Whiteside's dominance in this insular, increasingly crowded world of the Newton scholars was, by all accounts, unchallenged and not a little fearful. It was, Richard Westfall wrote in a 1976 paper on the "Newtonian Industry," a "claustrophobic and neurotic society, the members of which circle each other warily, jockeying for position, each of the others with one eye cocked on the Lord High Executioner, D. T. Whiteside." The man wielding the power, Westfall suggested, "stands at the center, beta function in one hand, citation from *Add. MS 3965.12* f. 186 *recto* in the other, striving with indifferent success to maintain law and order."[20] *Add. MS* is the Cambridge University Library's catch-all category for "additional manuscripts" in its holdings, a term that describes its accumulation of documents. The "beta function" stood for mathematical, the *Add. MS* for scribal, authority. And Whiteside had both. The precise folio that Westfall imaginatively placed in his hand was from Newton's notes on the *Principia*, the heart of his scientific legacy and a clue, once again, to his working method.

In 1841, in response to Halliwell's first publication for the ill-fated Historical Society of Science, Augustus De Morgan had wondered if there would ever be a paleographer who was also a mathematician, "with enough energy and leisure both to work the ore and the metal."[21] It was a rare man who could treat of mathematics and the study of ancient writing with equal skill and enthusiasm. More than one hundred years after De Morgan had asked the question, Whiteside, clutching the manuscripts both literally and figuratively, gave the answer. Born

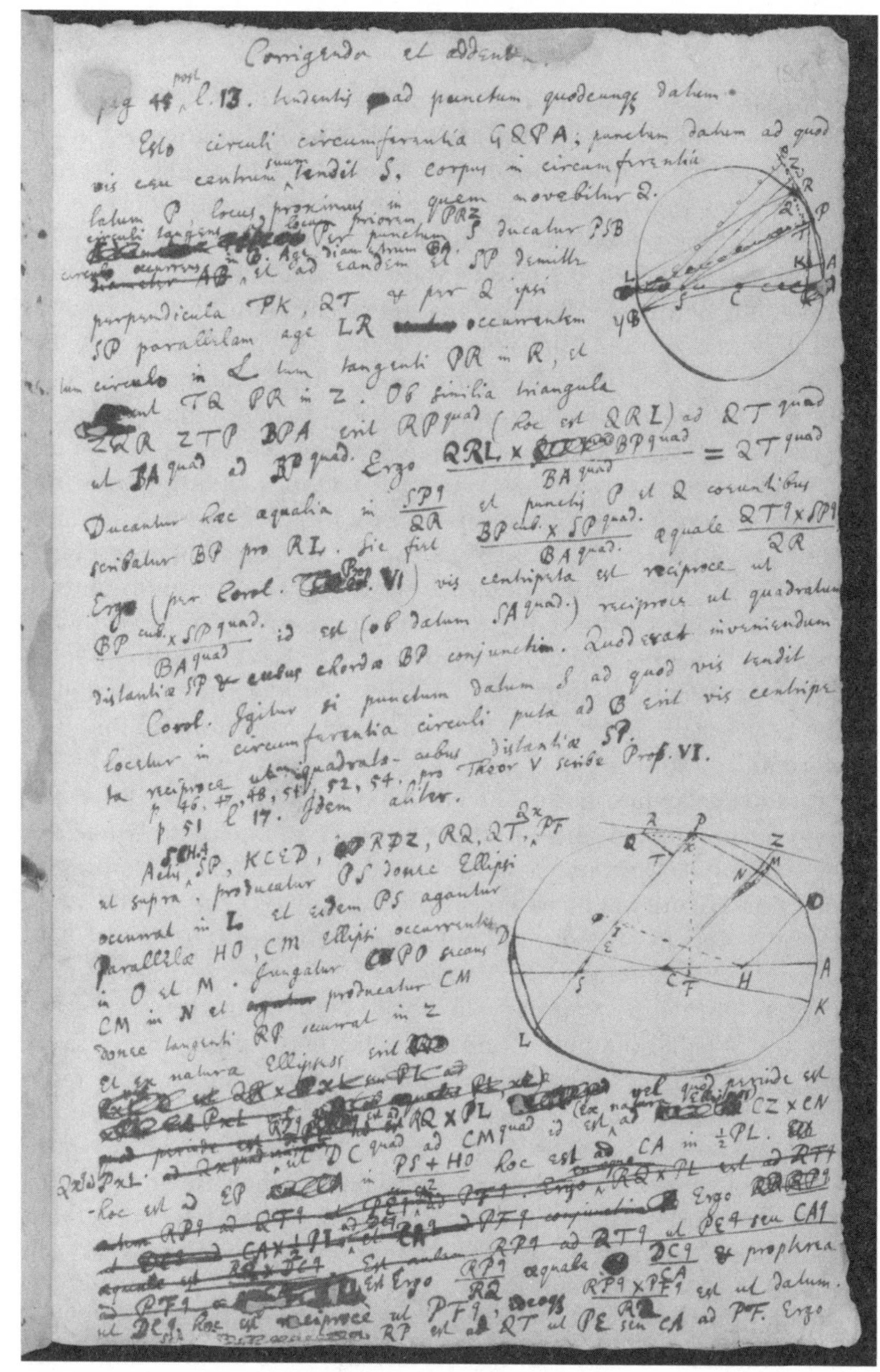

The folio of Newton's notes on the *Principia* which Richard Westfall satirically imagined in the hands of D. T. Whiteside, editor of *The Mathematical Papers of Isaac Newton* and the "Lord High Executioner" of the Newton industry. Deciphering Newton's method was a preoccupation of many scholars. MS Add. 3965 f186r. Reproduced by kind permission of the Syndics of Cambridge University Library.

to a working-class family in Blackpool in 1932, Whiteside had lost his mother as a child but made it through Blackpool Grammar School to Bristol University on a scholarship. There he had studied French, Latin, and mathematics. After a brief period of National Service, he had enrolled for a doctorate at Cambridge, almost on a whim, to work on a dissertation titled "Patterns of Mathematical Thought in the Later Seventeenth Century."[22] While doing the research for his thesis in May 1958, he had asked the librarian, also on a whim, if the university had any Newton manuscripts. The librarian produced the long-neglected scientific portion of the Portsmouth papers that Adams and Stokes had catalogued more than seventy years earlier. So began the project that would, twenty-three years and eight volumes later, be known as *The Mathematical Papers of Isaac Newton.*[23] By 1981 Whiteside had finally produced what so many had yearned for and despaired of ever seeing: an edition of mathematical depth, archival rigor, and comprehensive scope. Whiteside's achievement was unanimously recognized as the pinnacle of Newton studies up to that time.[24] It was only the mathematical papers, of course, and therefore just a shard of the broken glass that was Newton's archive, but who would say it was not also the most important?

Paradoxically the rigor of the edition, what Westfall called the "uncompromising technicality of the editorial commentary"—the very thing that the Newton papers required, after all—made it daunting to those who were not willing to engage directly with difficult mathematics. The edition repaid careful study, but it was not designed to convey Newton's work to a general audience. For some in the Newton industry, clarity was more important than the details of mathematical technique. It was not, Westfall reminded, enough to be merely good, or even very good, at mathematics. This stuff was difficult enough to require quite advanced training. Mathematicians were, frankly, unlikely to be interested in the obscurities of history. And even historians of science—even historians of Newton!—who were quite skilled in their treatment of his mathematics and science could find it hard going. Westfall guessed that this was not incidental. It was impossible to shake the feeling that "the technical level is deliberately set as an act of defiance."[25]

Westfall may have overstated the case, since rather than repelling study, Whiteside's edition laid the foundation for modern studies of Newton's mathematics.[26] But this paradox is nevertheless well-suited to Newton, who wanted it both ways: to seize priority and to communicate only with those select few who were prepared (in both senses of the word) to follow his reasoning. The question remains whether

D. T. Whiteside receiving the first Koyré Medal from the International Academy of the History of Science in 1968 for the first two volumes of *The Mathematical Papers of Isaac Newton.* Left to right: a senior printer at Cambridge University Press, Michael Hoskin, Joseph Needham, D. T. Whiteside, Brooke Crutchley (the university printer), Lord Adrian (chancellor of the university). Courtesy of Michael Hoskin.

Whiteside, who died in 2008, was deliberately imitating Newton, who had famously crafted a difficult work to put off "little Smatterers in Mathematicks." Does an edition need to reproduce its subject to such a fine extent? Whiteside may have said so. Others, Westfall among them, thought not. Whiteside's Newton offered neither justification nor apology. This was Newton the mathematical physicist, whose accomplishments had secured his own fame and fortune during his lifetime and a halo of adulation ever since.

Increasingly, however, the mathematical and scientific Newton was coming face to face with other Newtons, whose passions and proclivities could still embarrass or irritate or confound. The manuscripts had finally spoken, and the result was not constructive but deconstructive. The Newton of old was no more, and had not really been tenable even since De Morgan; the new Newton was not one man but many. As Westfall put it in 1976, "The Newton of the textbooks has come unglued. As yet, no one has re-assembled the pieces into a coherent form."[27] The poignant fact was that Whiteside's perfect edition, what Westfall had called the "premier edition of scientific papers of all time," had arrived at precisely the moment when it seemed no longer sufficient to the assorted members of the Newton industry. Pressing and exciting work was being done not only on the Newton of old, the familiar genius of mathematics and physics, but on a new Newton. This new Newton was not a marble monument but a shapeshifter. A new identity was revealed with each new publication. As more scholars studied the papers, more Newtons appeared, often bearing an uncanny resemblance to the scholars who found them.

Much of this work was energized by recent scholarship by Frances Yates showing that far from being limited to the Middle Ages, magical and mystical thinking (based on the writings allegedly composed by the great magus Hermes Trismegistus) had played an important role in science and philosophy well into the Renaissance.[28] The publication of volume 3 of the Newton correspondence in 1961, covering the years 1686–94, revealed that Newton had been preoccupied by alchemical and theological matters during the same years that he was composing the *Principia.* Inspired by Yates and empowered by the new documentation, scholars set about describing the many parts of Newton. Ted McGuire and Piyo Rattansi used unpublished sources to show that Newton was intensely concerned with the secret tradition of the *prisca sapientia,* whereby an elite cadre safeguarded ancient knowledge through the centuries.[29] McGuire, working alone, showed Newton's metaphysics to be philosophically much more engaged and more complicated than his published works, the *Principia* and the *Opticks,* indicated.[30] David Kubrin, drawing on theological manuscripts in the Keynes collection, described Newton's vision of a cyclical cosmos in which comets provided the divine mechanism for the renewal of dissipated energy.[31] Betty Jo Dobbs, one of the few women present in this coterie, made alchemy a subject that could no longer be disregarded in relation to Newton.[32]

From alchemy it was but a tripping step along the garden path to Newton's religious beliefs and personal eccentricities. Frank Manuel provided a set of biographical investigations, one on Newton as a historian (which relied mainly on the Keynes collection and manuscripts in New College, Oxford) and one on his personality, which drew on psychoanalysis to reexamine certain manuscripts such as, notoriously, a vocabulary list that Newton had written down in 1662.[33] Manuel demonstrated that Newton had gone off-piste, as it were, from the prescribed list and included his own, often highly charged choices that could be interpreted to reveal his feelings about his mother, his stepfather, and his religion. Manuel was also the first since Castillejo to dip into the manuscripts in Jerusalem and, in a third book, was the first professional scholar to engage seriously with the content of Newton's theological writing.[34] Westfall himself began work on what would prove to be the fullest biographical treatment of Newton since Brewster's Victorian books, one in which manuscript evidence was given central and unapologetic importance.[35] Even Newton's library, long neglected, received full scholarly treatment. John Harrison told the twisting tale of the 2,100 books in Newton's possession on his death, describing which of the nine hundred books that had survived had annotations in Newton's hand (his 1660 English Bible was the most heavily marked) and explaining Newton's idiosyncratic habit of dog-earing his books, often to mark instances where his own name appeared.[36]

There was ample reason for this profusion. Newton was, after all, a scholar's scholar. As A. R. Hall put it, "Has any man ever read, annotated, studied more than Newton?"[37] The marks of Newton the scholar, laid down more than three hundred years earlier, persisted with remarkable vividness. As a first-year undergraduate at Cambridge, Newton had recorded his expenses for such tools of scholarship as a "paper booke," a "quarte bottle" and "inke to fill it," as well as a brush, paper and quills, a pound of candles, and a lock for his desk.[38] Though he sometimes bought them, Newton liked to mix his own inks for writing. Made by grinding the bulbous black galls of oak trees with gum Arabic and iron sulfate, they were idiosyncratic and long-lasting. Recording a recipe for what he called "excellent" ink, he took pride in noting, "With this ink new made I wrote this."[39] (See frontispiece.) Some 350 years later, the ink remains clear and dark.

The practical habits of reading, writing, and note taking that he acquired early on with the use of those "paper bookes" and quills were fundamental to his productions in all of his myriad areas of interest. By understanding Newton as a scholar, pen in hand, amid his books and papers, it becomes easier to see him as a single, if singular man at work. Though his interests ranged widely, his scholarly habits provide a thread with which to knit them to one another. They also provide an intimate link to the man himself, demonstrating the way he lived his life on a daily and even hourly basis. As artifacts of thinking, in all its many forms, his papers document the progress of his mind. They reveal Newton—a taciturn and largely solitary man—conversing with a huge array of authors. In the world of paper, Newton is both garrulous and appreciative, respectful and probing, impassioned and impatient.

Alongside this refreshed awareness of Newton's life as a writer of texts came a poignant sense of the man's social isolation. "With every volume of the *Papers,*" wrote Michael Mahoney in a review of volume 8 of Whiteside's *Mathematical Papers,* "the tragedy of Newton's isolation stands forth more clearly."[40] Mahoney lamented that no students had picked up where Newton left off and that the great wealth of Newton's knowledge had therefore remained buried in his papers, which had been inherited by people who did not appreciate their contents, only that Newton had written them.

Just how limited Newton's social world was, and how meaningful his degree of separation, is not a simple question to answer, though much ink has been spilled on such a debate. Over the centuries much has been staked on the image of Newton's solitude, which supposedly allowed him to remain detached not simply from the cavils of other philosophers but from the material world around him—the world that was building its way toward an industrial revolution. Newton's relationship to that revolution, to the machines and manufacture that transformed first Britain and then the world, has been a problem for many in the (ironically named) Newton industry. Eagerness to preserve the image of Newton the solitary thinker has persisted despite, among other things, his dependence on the very practical observations of men such as Flamsteed that he needed to build his *Principia* (though it is no easy matter to link either the *Principia* or the *Opticks* to the material changes wrought by the Industrial Revolution).[41] Another difficulty with this image of solitude is that Newton was not friendless. In Cambridge his secretary reported that even during the 1680s, when he was deeply engaged in the making of the *Principia,* Newton had two or three close friends, including the chemist Giovanni Vigani, "in whose company he took much delight and pleasure," though he did abruptly break off the friendship after Vigani told a "loose story about a Nun."[42] For the last thirty years of his life, Newton lived in the intensely social, even political world of administrative London, hosting parties at his house with the help of his young half-niece Catherine and engaging deeply in the business of the Mint. Nevertheless it is clear that Newton often found social interaction trying and an unwelcome distraction from his studies. His social engagement with his own contemporaries could not match the richness of his inner world.

For historians working directly with the treasures of Newton's private world, it seemed natural to stress the depth and intensity of this personal realm. The sheer abundance of the papers was obvious to anyone who had even passing knowledge of them. More substantively those who read and studied the papers soon realized what excesses they contained. There was a surplus of subject matter, the many shards that alternately caused delight and frustration in various scholars, and, even deeper, a surplus of imaginative detail within each of the many genres in which Newton worked. As he sat alone at his desk, quill and paper to hand, he conversed with the Vandals and Visigoths of the fourth century; with allegorical figures from alchemical texts, such as "Neptune's Trident," the "Green Lyon," and "Mercury's Caducean Rod"; with the phantasmagorical monsters, horsemen, whores, and queens in the books of Daniel and Revelation; and with the mathematical models that populated his imagination.

Newton's ability to move among such disparate intellectual realms may be best understood as an aspect of his faith. He believed in a world superintended by a divine omnipotence in which myriad forms of life, unimaginable to mere human beings, were possible. The extreme variety of Nature was ever present. It was enough to look through a magnifying glass at the creatures in a drop of water to grasp the "strange & wonderful nature of life & the frame of Animals" on earth; those in heaven would be more wondrous still. Newton had revealed the invariant mathematical rules that governed the motions of both terrestrial and celestial objects, yet he took comfort in this extreme variability. The power of the divine extended beyond the marvels of form to those of movement. He noted the "sufficient power of self motion" that certain higher beings, such as angels, had to move "where they will, & continue in any regions of the heavens whatever, there to enjoy the society of one another." This mobile social order of the heavens was, in Newton's opinion, a freer and therefore more desirable place than any fixed habitation could be. To "have thus the liberty & dominion of the whole heavens & the choise of the happiest places for abode seems a greater happiness then to be confined to any one place whatever."[43] Newton re-created this happiness for himself in the realm of his imagination, roaming at will across whatever subject, whatever genre he chose.

It makes sense that it would require an industry, and many generations, to come to terms with this sort of man. The Newton scholars, as variable and strange as the creatures to be seen in a drop of water, matched the variety of Newton's own thought. Drawn to different aspects of the man revealed by his papers, the scholars betrayed their own predilections. For Keynes, who knew what it was to live one life publicly and another in private, Newton was a premodern magus in the body of a rationalist. For Yahuda, he was a student of history and of language who applied equal parts passion and system to his quest to derive a singular truth from a palimpsest of ancient words. For Cohen, Newton was a mathematical physicist, the author of the masterwork *Principia*, and the possessor of a distinctive, "Newtonian" style, which involved repeatedly testing mathematical models of the world against the physical universe. For Dobbs, he was an alchemist who considered the earth a living, breathing organism. Manuel, perceiving the world in psychological terms, thought Newton's emotional relationships were primary, that with his mother above all. In the end Newton got the scholars he required, and his long-neglected papers enabled them to find what they wished for, as they trailed Newton across the wide open spaces of his mind.

14

# The Search for Unity

Having dived down the rabbit hole into the warren of Newton's inner world in their attempt to piece together the mess of his manuscripts, scholars struggled to make it back into the clear light of day. The question was almost irresistible: What did it all add up to? Banal as it was, this question begged a certain kind of answer, an answer that put Newton back together again.

Unity became the great white whale of this Newton industry—the elusive, desirable prize. As the papers revealed new facets of Newton, scholars tried to fuse the man back together, to recover the singular consciousness that cast its glow across such disparate subjects. For those who sought to rehabilitate the alchemy and the theology, showing the connections between the science and the nonscience buoyed the whole structure. Unity kept Newton out of the muck of heresy and mysticism, tethered his wilder speculations to an overarching plan, a rational structure of thought in which the great mind sought to place every aspect of human endeavor. That Newton's own embrace of the freedom to move at will might militate against such a reconstruction did not seem to register with the community of scholars.

This emphasis on unity was due in part to the rise of psychobiography. An avid and self-confessed practitioner of this historical art, Frank Manuel used the phrase "overwhelming desire" to describe Newton's quest to find the hidden order in the world. "If nature was consonant with itself," Manuel wrote, working to reconcile Newton's theological and scientific undertakings, "so was Isaac Newton's mind." Manuel argued that Newton had "a compelling drive to find order and design in what appeared to be chaos, to distil from a vast, inchoate mass of materials a few basic principles that would embrace the whole and define the relationships of its component parts."[1] For Manuel, at least, Newton's mental unity relied on an underlying psychic need, a "drive," that was based in his belief that God's will was made manifest in all his works, scriptural and natural.

Even those scholars without an explicitly psychological approach were vocal in advocating for Newton's singular mental worldview as an explanation for his varied interests. Picking up on the theme of unity in her book on Newton's alchemy, Betty Jo Dobbs considered faith to be the conceptual glue that bound Newton's thought together. Truth was, quite literally, divine: "The Janus-like faces of Isaac Newton were after all the production of a single mind, and their very bifurcation may be more of a modern optical illusion than an actuality.... In Newton's conviction of the unity of Truth and its ultimate source in the divine one may find the fountainhead of all his diverse studies."[2] In pushing the pendulum back against scholars who had long sought to excise the alchemy and the religion from Newton's public image, scholars like Dobbs and Manuel worked hard (some would have said too hard) to show direct connections between the different elements of Newton's archive.

Published some twelve years after he had first encountered the Jerusalem papers, David Castillejo's 1981 *Expanding Force* continued the trend. Castillejo's book was the first to bring material from Newton's manuscripts on alchemy, prophecy, and chronology together in one analysis.[3] Castillejo was a unifier par excellence, and his approach was simple: "to show that the whole of his unpublished and published work—his mechanics, optics, alchemy, and church history—does in fact form a single body of thought." He argued that Newton had tried to concoct a universal expanding force, an inversion of gravity. This force extended to history as well as physics, thereby connecting Newton's chronological with his natural philosophical writings. The "expanding law or force" was evident in the radiation of light, in chemistry, in biological growth, and even in the "mind and behavior of human beings."[4] Castillejo applied numerological analysis to Newton's writings, identifying such a profusion of overlapping and coordinated numbering systems that it is almost impossible to follow him. The numerology was the basis for Castillejo's argument on the unity of Newton's thought: the numbers match up, he claimed. A simple example of the correspondence Castillejo found is between the four-sided ray of light that Newton posited and the four-sided Temple of Solomon whose size he calculated; many of the connections were harder to trace.

Like many who believe they have uncovered secret meanings in coded texts, Castillejo argued that Newton had worked hard to obscure his meaning in his texts: "So at first, a Newtonian text appears to say almost nothing; but if we search clues concerning the monument he is shaping, and, without forcing nor manipulating his statements into wishful patterns of our own, merely let them float in the air without touching them, then Newton gradually begins to reveal himself, and his texts grow simpler and simpler until they become transparent, exhibiting huge, gigantic structures." Like Sampson, Castillejo emphasized the seamlessness of the archive. While topics appeared initially distinct, gradually it became clear that "his whole thought is not only very simple, but also absolutely, ruthlessly inter-locked."[5] On their own the extensive quotations from unpublished Newton material in Castillejo's book made it a potentially valuable resource for Newton scholars. But the unusual book was largely unread and sparingly reviewed.[6] Castillejo—or his publisher—did himself no favors by advertising on the back cover an upcoming book on consciousness, detailing Castillejo's ability to "register perceptible pressures on his brain akin to headaches, which coincide with the varying breathing patterns of those around him." Using over twenty years of observation, Castillejo had managed to establish that "a continuous non-telepathic communication is taking place between minds... and animals are also involved."[7]

For those who studied the traditional triumvirate of optics, mathematics, and natural philosophy, the unity of Newton was less pressing and quite possibly harbored problematic implications for the purity of their subject, similar to those with which Brewster had grappled years earlier. They did not seek to devolve or deconstruct their man; they were content to use the new trove of scientific material to continue to ask questions about method rather than psychology or metaphysics. Some chose to remain quiet on the subjects about which they knew, or preferred to know, less. Whiteside was unapologetic about his ignorance of much of Newton's nonscientific writing. Westfall too admitted that his biography was primarily a "scientific" one, and though he consulted some of the theological and alchemical writings, his Newton remained, much like Brewster's, a man of unsurpassed genius. Westfall had immersed himself as deeply in the archive as anyone, and yet after so many years studying him, in the end Newton remained "wholly other," a genius whose mind was not accessible to other men. But in his otherness, Newton was closest to another abstract noun: truth. Describing Newton's scientific coming of age in the 1670s, Westfall writes,

"For eight years he had locked himself in a remorseless struggle with Truth. Genius of Newton's order exacts a toll. Eight years of uneaten meals and sleepless nights, eight years of continued ecstasy as he faced Truth directly on grounds hitherto unknown to the human spirit, took its further toll."[8]

These yearnings toward capitalized notions of Unity and Truth now betray, in retrospect, a somewhat embarrassing lack of scholarly decorum, a blush of enthusiasm that does not hold up with the passing of time. The impulse to simplify Newton came from both sides of the aisle: from those who wished to reveal a Newton of many parts as well as those who focused simply on the science. Both approaches missed the more open-ended, less concretized possibilities that the rehabilitated archive offered. Rather than revealing the substructure upon which Newton built his philosophy or the process of his most rational thought, the very looseness and foulness of the archive, its population of drafts and redrafts, reveals a mind always changing. ("Never at rest" was the phrase Westfall plucked from a letter to use as his title.)[9] A mind always in motion may indeed be seeking both truth and unity, but it doesn't seem satisfactory to suggest they can be so easily located. The very shape and contents of the archive suggest that for Newton there was never a single avenue toward knowledge, nor a single form in which it might be captured. The deep and substantive revisions to his major published works, the *Principia* and the *Opticks*, and the overwhelming evidence of continuous revision in his unpublished manuscripts confirm an image of Newton as a man almost continuously at work and just as continuously at revision.

There were practical problems with trying to capture this dynamism of the archive. Whiteside had acknowledged the "violence" of the editing and publishing process, whereby irregularities and ambiguities were smoothed out or even erased. It was a regrettable but also, Whiteside concluded, a necessary feature of making the material more widely available. To provide a proper edition, hard choices had to be made, rough cuts hewn in the granite of Newton's seamless archive. And there were always critics of the choices made by editors. Ivor Grattan-Guinness, himself a historian of mathematics, performed the editorial equivalent of an autopsy on Whiteside's treatment of a single folio of Newton manuscript.[10] Working from a facsimile of the manuscript, he demonstrated that Whiteside had decided in his printed edition to include some and exclude other snippets of material that had, manifestly, lain side by side on the manuscript page. Whiteside, according to Grattan-Guinness, had done the unforgiveable and "played Newton in the sense of taking a manuscript, or a related group of them, to some next stage of preparation."[11] *Playing Newton*. It was the ultimate transgression of the editor. And yet, was that not also his most necessary task? An editor could grasp the material only by assuming the thought practices of his subject. Where, exactly, was the line that divided the necessary leap of sympathetic imagination from that of overweening vanity, presuming to think for Newton? This was the conundrum for the editor, who must inhabit his subject's mind without daring to speak for him. Matters were not helped by how predictably meager were the rewards of this kind of editorial abasement.

Ultimately—as generations of scholars discovered—Newton's papers will always remain unfixed and unstable. The manuscripts abound in revisions. More than any other comparable figure from the time, Newton was his own lifelong editor. He was always there first, revising, deleting, and redrafting across a lifetime of study. He constantly inspected his own material in his search for deeper meaning within it. This was both an effect and a cause of his lifelong aversion to publishing. By not publishing, Newton made it possible to continually revise what he wrote. Even publication did not mark the end of the process, as the extensive revisions between consecutive editions of the *Principia* and the *Opticks* indicate, but publication made the process more, well, public and querulous and generally time-consuming in the way Newton had forsworn, despairing of that "litigious Lady," philosophy. Newton's practice of continuous revision makes it very hard—perhaps impossible—to grasp the meaning of his manuscripts because there is no fixed meaning to find. Add to this the devilish difficulty of fixing dates to so much loose material with only infrequent references to contemporary events, and the problem becomes magnified. Time and again an appetite for suffering appears among the list of qualifications for would-be editors of Newton's papers.

Only by embracing the dynamic nature of these papers, for the ways they record the active seeking in which Newton was engaged, do we see them for what they are: not a single story but a mass of intersecting stories.[12] This kind of thinking veers into uneasy territory, where instability trumps fixed meaning and multiplicity beats singularity. It is an uncomfortable place to be. For some, it is also the only honest ground on which to stand.

# Epilogue: The Ultimate Value

As time passes, the chance that additional, unknown Newton papers will emerge decreases but never falls to zero. There is always the possibility that someone will open an attic or a forgotten closet, and a stash of old paper will come to light.

That happened at Maggs Bros. a few years ago, when Arnold Hunt found a leftover Newton lot from the 1936 Sotheby's sale "gathering dust in a cupboard (and given the amount of dust I think I must have been the first person to set eyes on it for at least fifty years)."[1] The manuscript, part of the original Lot 324 purchased by Maggs in 1936 and containing information on the longitude problem, was promptly identified, catalogued, and sold for an unspecific amount to an English collector. More recently still, another missing Sotheby lot turned up at the Maggs offices, this a copy that Keynes had taken on approval and then returned. Such pages can prove very valuable. In one recent auction three folio pages in Newton's hand on "the question stated about abstaining from blood" dated to circa 1719 were sold by Bonham's for £85,000. Manuscripts on subjects more directly linked to Newton's primary scientific achievements go for even more. A set of four pages of Newton's handwritten notes in Latin and Greek comparing classical views of gravity sold for £170,000 by Christie's in 2007. A cache of Newton-related material also made it into the marketplace nearly one hundred years ago, in 1929, when much of Newton's library was unwittingly sold at auction for just £170 (less than half of what it was valued in 1727 and approximately £5,000 today) by a man who didn't realize they formed part of his own library.[2] Those books were scattered to many buyers who were as unaware of their provenance as the seller. One man who did realize their true importance was Heinrich Zeitlinger, who bought as many as he could, many of which were eventually purchased by a charity called The Pilgrim's Trust and donated to Trinity College, Cambridge, in 1943.[3]

More spectacularly a treasure trove of rare manuscripts emerged on the public market in 2000, when a family feud led the present Earl of Macclesfield to sell portions of his family's library (among many other possessions). The earl had inherited the library from his grandfather nearly forty years earlier but had been unaware of precisely what it contained: an archive of famous but largely unexamined Newton papers that were shambolically housed in a "cupboard in the library, along with a mass of other papers."[4] It was an "intellectual time capsule in a house that is itself a time capsule,"[5] a snapshot of life in 1750 that had remained nearly unexamined for 250 years (save for publication of excerpts by Stephen Rigaud). The cupboard must have been a large one: in all five hundred notebooks and an equal number of loose manuscript folios were recovered from their shabby accommodation and eventually sold by the earl to Cambridge University Library. That the sale price of £6.37 million was considered "fair" indicates just how far the market for Newton manuscripts had come since the Sotheby's sale.

Sometimes the forces of history conspire to accrete rather than, as is more usually the case, to disperse material. The Macclesfield collection is such a collection; it is actually a series of nested collections. At its heart are letters written to or by or passed via a man named John Collins that make it possible to listen in on the lively, contentious mathematical conversation that flourished in Britain and the Continent in the 1660s, 1670s, and 1680s. Seventeen years Newton's senior, Collins was the mathematical point man for Royal Society correspondence in its earliest years, providing expert guidance to the Society's secretary, Henry Oldenburg, who himself served as the hub for an ever widening network of mathematicians, natural philosophers, and savants in Britain and on the Continent. Collins was the man, as De Morgan put it, who "wrote to everybody, heard from everybody, and sent copies of everybody's letter to everybody else."[6] Though not himself highly skilled as a practitioner, he knew excellence when he saw it. Collins met Newton in November 1669, when Newton was just a twenty-six-year-old Cambridge fellow, practically unknown outside of the university. Collins immediately began a correspondence with him, and it was to Collins that Newton wrote about the development of his mathematical researches.

Collins was intimately involved with the exchange of letters that led Leibniz to learn about Newton's work on infinite series and his method of determining tangents to complex curves. It was with Collins that Leibniz met in October 1676 in London, and Collins who showed him copies of Newton's papers. It was Collins who then shared Leibniz's letters with Newton in 1677. Collins earned a place for himself in the third edition of the *Principia* with a reference to a letter from Newton to Collins of December 10, 1672.

The Collins correspondence ended up in the Macclesfield library via a somewhat complicated path. After Collins's death, they were inherited in 1708 by a man named William Jones, who served as a tutor to both the first and second earls of Macclesfield. For Newton, the timing of this inheritance was good. Collins's papers, made accessible to him by Jones, proved very useful in his quest to prove priority in the matter of the calculus. In 1712 Jones provided the papers available to the Royal Society committee (controlled by Newton) that had been assembled to adjudicate the question of priority for the calculus, and many of the papers were incorporated in the official report of that committee, the *Commercium Epistolicum*. When Jones himself died, both his and Collins's papers were inherited by the Second Earl of Macclesfield and were to remain in the family's possession for the next two and a half centuries, when the library was finally put up for sale.

It is impossible to know precisely what other treasures may be hiding in the cupboards of old British homes. Nonetheless it is extremely unlikely that anything on the scale of the Macclesfield collection will ever surface again. There are very few remaining great libraries that remain intact and fewer still containing anything like the quality and importance of these Newton materials. The churn of books and manuscripts that Munby witnessed is unlikely to recur.

How much is a piece of old paper worth? To be more precise, how much is a piece of paper that Newton wrote on worth? It is hard to say. A manuscript that concerns mathematics or physics will garner more interest, and value, than one on which the great man struggles with a piece of Church history. All things being equal, a clean copy of a manuscript is worth more than a ragged one, a large manuscript more than a small one, a rarity more than a commonplace. The only sure answer, however, is that a manuscript is worth what someone is willing to pay for it. This maddening reply is one that book dealers supply, with a Cheshire Cat smile, to indicate the capriciousness at the heart of their business, the way that passion rather than reason finally rules the market.

It is a strange experience for the uninitiated to walk around a rare books and manuscripts fair. The manuscripts in particular seem so exposed, lying out for the passing public to consider, to handle (carefully), and, for the right price, to purchase. It is easy to bemoan the private collector as a selfish and somehow artless soul who acquires the ideas of men as though they were things and keeps them locked up, out of the reach of those who (the implication goes) might truly appreciate them. Yet it is the collectors who are both willing and able to deliver on the value assigned to the manuscripts. Collectors and, as we've seen in the case of the Sotheby sale, their dealers play an important role in identifying documents that might otherwise be sold into obscurity and preserving them until an appropriate institution can raise either the funds or the awareness necessary to procure them. Having paid what is in some cases an extraordinary amount of money to acquire a manuscript, the collector will almost certainly take good care of it. That it may represent an investment is all the more reason for it to be carefully preserved. Because of this, and because individual lives are limited but institutions can endure, valuable books and manuscripts have tended, throughout the course of the past hundred years or so, to end up in institutions such as libraries, universities, and research institutions. The paper itself, if it is protected from damage, loss, and theft, has a life of its own that is longer than the life of any one collector, any single institution.

The great majority of the Newton papers that have survived the nearly three hundred years since the death of their author have landed in research libraries. Though they have taken variously circuitous routes to end up back in the place where many of them were drafted, about half of Newton's surviving papers are in Cambridge. The lion's share by far belongs to the University Library, which owns the scientific papers donated by the Earl of Portsmouth in 1888. These include among other treasures Newton's annotated copies of the first and second editions of the *Principia*, his early research program titled "Certain philosophical questions," and the notebook he inherited from his stepfather in which he made extensive and important mathematical notes. Trinity College has one of Newton's undergraduate notebooks (in which he kept expenses) and a series of important letters he wrote to Richard Bentley. King's College, to which Keynes left his collection, is home to most of Newton's alchemical papers, including his vast bibliographic summary of alchemical texts, the *Index Chemicus*, as well as a fair number of theological writings purchased by Keynes at the sale. Also in Cambridge, the Fitzwilliam Museum owns the notebook in which Newton recorded the sins and expenses of his late adolescence and early adulthood.

Outside of Cambridge, the largest repository of Newton material in England is the National Archives at Kew, which holds all of the papers relating to Newton's tenure at the Mint, amounting to some eight hundred distinct items that have not yet been thoroughly examined by historians. Significant manuscripts are also in New College, Oxford, which holds the theological writings that Catherine Conduitt had hoped the Reverend Ekins might help to get published. The National Library of Israel (as of 2008 the new name for the Jewish National and University Library) in Jerusalem holds Yahuda's large collection of Newton's theological papers. In America, the Huntington Library now houses the Grace K. Babson Collection of the Works of Sir Isaac Newton, rich in books and possessed of a few manuscripts.

Finally, dozens of individual manuscripts, mostly those scattered by the Sotheby's sale, can be found in libraries around the globe, including St. Andrews University Library in Scotland, the Fondation Martin Bodmer in Geneva, the University of Texas at Austin, the Massachusetts Institute of Technology, the William Andrews Clark Memorial Library in Los Angeles, Stanford University, and the Smithsonian Institution, home to the Dibner Library, which contains eight of Newton's alchemical manuscripts. Roughly thirty different manuscripts that were sold at the Sotheby's sale are currently held in unknown locations by unknown buyers or may have been lost.

The list of locations is dizzying, though, remarkably, most of the material scattered so widely by the Sotheby's sale has been carefully catalogued, and the majority of the most important documents are in a handful of major libraries. The dispersal of the papers is inconvenient, not devastating. And the inconvenience of their separation is shrinking to a vanishingly small point, as physical location is less and less important to scholars. Today the Newton papers can increasingly be located and studied at a few significant digital addresses, to which more and more material is being added every day.

Of these, the oldest and most extensive is the Newton Project, which has transcribed and made publicly available online all of Newton's religious writings along with the most significant of the optical, physical, and mathematical writings and his correspondence. A related website, the Chymistry of Isaac Newton Project, is doing the same for his alchemical notebooks, along with a project to re-create many of his alchemical experiments, while the Cambridge University Library has made digital images of many of the scientific papers it holds available online.

These websites often include facsimiles (digital images of the papers), but just as important, if not more, are the searchable transcriptions that are provided. These transcriptions represent a further step toward editorial transparency by encoding and making available to readers information about where material has been added or deleted by Newton. Along with the ability to easily search for keywords, this so-called diplomatic transcription renders the manuscripts accessible in a fundamentally new way while preserving some, if not all, of the ambiguities of a manuscript. Today all of Newton's private writings on religious and alchemical topics, some 7 million words, are now freely available online. In a radical fashion, what was once private is now utterly public.

Something about having all that material available makes it clear that complexity trumps simplicity, that there are devils in the details worth attending to, and that both questions and answers may come from unexpected corners. The Internet, after all, welcomes readers who may not have the credentials that would admit them to the leading research institutions. The conversation is potentially much more clamorous, unregulated, and wider than ever before. Professional scholars have begun to make use of these materials in their new, digitally accessible form. Today the pendulum of scholarly debate about Newton's life and work has swung once again, away from conceptions of "unified thought" and back to something like where things stood during Newton's own lifetime, when his natural philosophy was considered to be largely separate from his theology.[7] Nevertheless the current vogue for a Newton of many parts probably tells us as much about our contemporary fixations—on multiplicities of meaning, on diversity, and on the idea that interpretation is a never-ending endeavor—as it does about Newton himself.

The task, for those who choose it, is to reimagine Newton in terms that are both historically resonant and relevant to today's world of scholarship. The online Newton is joined by the digital avatars of his contemporaries, men such as the poet John Milton, the philosopher John Locke, the natural philosophers Robert Boyle and Robert Hooke, and the mathematician Gottfried Leibniz. For many scholars, the excitement lies not in unfolding once more Newton's personal origami but in understanding Newton as a man among his peers.

Meanwhile stories about Newton's secrets continue to emerge. We seem to have an unlimited appetite for tales that confound the image of the perfect rational thinker with tales of his predilection for theological and other pastimes. People continue to remark upon Newton's alchemical obsessions, his religious heresies, and his period of insanity. While these stories can usefully complicate our vision of Newton, and of science more generally, they also play a role in confirming a narrow vision of science. To be surprised by Newton's range of insights requires that we maintain an underlying belief that scientists "should" be one way: rational, focused, consistent. That we continue to be surprised to discover that scientists are, after all, complex human creatures like the rest of us may owe more than we realize to the way that Newton's life has been told and retold through the centuries.

Gradually a sense of the "true" complexity of Newton (equivalent to the true complexity of any human) may trickle down to the general public. However, that may be a red herring in our quest for historical and therefore self-understanding. It may be more interesting, and more illuminating, to consider why so many of us continue to be surprised by Newton's heterodoxy of thought. Is there something nourishing in the seeming contradictions at the heart of our great scientific hero? Those who feel removed from the scientific culture that Newton begat may feel it is a way to claim back something of Newton for the regular man, who abounds in self-contradiction and private truths. Perhaps we are not quite ready to accept an image of the scientist as a complex man. This commitment to the scientist as saint reveals the power of hero myths—which both provoke and repel attack—as well as the enduring strength of the idea of the scientist-hero more specifically.

Because Newton has remained famous since his lifetime, the process of sorting out the contradictions in his life can no longer be considered a singularly personal affair. He has become a part of our wider culture. Our understanding of him will always be conditioned by our expectations about what a scientist is or should be, and these expectations have, in turn, been shaped by our understanding (or misunderstanding) of Newton's own life. There is a circularity here that may seem confusing but is instructive: in our successive versions of Newton we can see the traces of ourselves. In searching for the papers, and the real Newton within them, we are also, always, searching for ourselves.

This is not to say that we can never get close to Newton, that he recedes as in a hall of mirrors. The remarkable thing is that the bits of paper that Newton left, fragmentary and confused as they are, allow us to get as close to his mind as they do. Thanks to the men and women described in this story—who stashed the papers and pursued them, sold and purchased them, studied and struggled with them—we can approach the worldview of a man who died nearly three hundred years ago. That Newton was an exceptional man all who study him would agree. He was also a man of his time. The papers give us the opportunity to try to understand both aspects: the extraordinary and the ordinary Newton. We owe the simple act of writing this intimacy with the past. In among his drafts and revisions, we are as close to Newton as his pen was to paper.

# Notes

## Prologue

1. Robert Skidelsky, *John Maynard Keynes: The Economist as Saviour, 1920–1937* (New York: Trans-Atlantic, 1993), 217.

## Chapter 1

1. The biographical detail surrounding Newton's death is from Richard S. Westfall, *Never at Rest: A Biography of Isaac Newton* (Cambridge, UK: Cambridge University Press, 1980).
2. William Stukeley, "Revised Memoir of Isaac Newton," MS/142, Royal Society Library, London.
3. For a description of the inventory, see John Harrison, *The Library of Isaac Newton,* (Cambridge, UK: Cambridge University Press, 1978), 29. The inventory is in the Public Record Office, Kew, as "A true and perfect inventory of all and singular goods, chattels and credits of Sir Isaac Newton... taken and appraised on the 21st, 22nd, 24th, 25th, 26th, 27th, days of April... 1727" and was signed by John and Catherine Conduitt, PROB.3/26/66.
4. The full text of the inventory is printed in Richard De Villamil, *Newton: The Man* (London: G. D. Knox, 1931), 54–55.

## Chapter 2

1. Cited in Richard S. Westfall, *Never at Rest: A Biography of Isaac Newton* (Cambridge, UK: Cambridge University Press, 1980), 870.
2. John Conduitt, "Drafts of various sections of the Memoir of Newton," Keynes Ms. 129.02, 4r–v, King's College, Cambridge. (All Keynes manuscript material is held at King's College, Cambridge.)
3. Keynes MS 127a.5.
4. The quotation is from the codicil to Kitty Conduitt's will January 26, 1737, cited in full in D. Brewster, *Memoirs of the Life, Writings, and Discoveries of Sir Isaac Newton,* 2 vols. (Edinburgh: Thomas Constable, 1855), vol. 2, 341, n51.
5. Conduitt's copy of Pellet's list is Keynes MS 127a.4. Subsequent references here are drawn from another version published in Charles Hutton, *A Mathematical and Philosophical Dictionary... In Two Volumes* (London: J. Davis, 1795), vol. 2, 155–57.
6. Isaac Newton, *The chronology of ancient kingdoms amended. To which is prefix'd, A short chronicle from the first memory of things in Europe, to the conquest of Persia by Alexander the Great* (London, 1728).

7. Isaac Newton, *De mundi systemate liber Isaaci Newtoni* (London, 1728); Isaac Newton, *Observations upon the prophecies of Daniel, and the Apocalypse of St. John. In Two Parts* (London, 1733).
8. John Aubrey to Anthony Wood, February 23, 1674, Bodleian MS Wood F 39, f255, cited in Michael Hunter, *John Aubrey and the Realm of Learning* (London: Duckworth, 1975), 65.
9. Cited in Hunter, *Aubrey*, 65.
10. John Conduitt, "Fair Copy of the Memoir of Newton," Keynes Ms. 129.01.
11. Keynes MS 131a, 1.
12. Keynes MS 130.12.
13. Keynes MS 130.02, 1.
14. Keynes MS 130.02, 3.
15. Keynes MS 130.07.
16. Keynes MS 130.10, 1r.
17. Keynes MS 130.02, 20.
18. Keynes MS 130.02, 20.
19. Keynes MS 129.01, 8r.
20. Keynes MS 129.02, 4r.
21. Keynes MS 130.10, 3r.
22. Keynes MS 130.07, 3r.
23. Keynes MS 130.15, 1.
24. Keynes MS 130.02, 90r.

### Chapter 3

1. For which see William Whiston, *A Collection of Authentick Records* (London: 1728), 1070–82.
2. Mead to Stukeley, London, April 4, 1727, Bodleian MS. Eng. misc. c. 114, f. 50, cited in David Boyd Haycock, *William Stukeley: Science, Religion and Archaeology in Eighteenth-Century England* (Woodbridge, UK: Boydell, 2002), chapter 8, n57.
3. John Craig to John Conduitt, April 7, 1727, Keynes MS 132, f3.
4. R. Wodrow, *Analecta: or, Materials for a History of Remarkable Providences; Mostly Relating to Scotch Ministers and Christians*, 4 vols. (Edinburgh, 1842), vol. 3, 461–62.
5. John Conduitt, "Draft Memoir of the Life of Newton," Keynes MS 129.01, 12v.
6. Wodrow, *Analecta*, vol. 3, iv, 59.
7. See Scott Mandelbrote, "Eighteenth-Century Reactions to Newton's Anti-Trinitarianism," in *Newton and Newtonianism: New Studies*, edited by J. E. Force and S. Hutton (London: Kluwer, 2004), 96.
8. Isaac Newton, *Observations upon the prophecies of Daniel and the Apocalypse of St. John* (London, 1733), part 1, chapter 2.
9. Voltaire, *Letters concerning the English Nation* (London, 1733), 155.
10. Voltaire, *Letters*, 159.
11. Arthur Bedford, *Animadversions upon Sir Isaac Newton's Book, Intitled the Chronology of Ancient Kingdoms Amended* (London, 1728), 143. And see Arthur Bedford, *A Serious Remonstrance in behalf of the Christian Religion against the Horrid Blasphemies and Impieties which are still used in the English Playhouses* (London, 1719).
12. Bedford, *Animadversions*, 143.
13. Bedford, *Animadversions*, 2.

14. Waterland to Zachary Grey, February 5, 1735, British Library (hereafter BL), MS. Add. 5831, fols. 172r–3r, cited in Mandelbrote, "Eighteenth-Century Reactions," 99.
15. Cited in Scott Mandelbrote, "Newton and Eighteenth Century Christianity," in *Cambridge Companion to Newton*, edited by I .B. Cohen and G. Smith (Cambridge, UK: Cambridge University Press, 2002), 409.
16. For an extensive analysis of Newton's method and aims in the *Chronology* and how they relate to his scientific method, see Jed Z. Buchwald and Mordechai Feingold, *Newton and the Origin of Civilization* (Princeton, NJ: Princeton University Press, 2012).
17. Codicil to Catherine Conduitt's will, cited in full in Brewster, *Memoirs*, vol. 2, 341, n51.
18. The papers are held in the Bodleian Library, Oxford, as New College Ms. 361.1–4.
19. Barton to Hanbury, March 25, 1757, Trinity MS R. 16.38:415r, cited in Whiteside, *Mathematical Papers*, vol. 1, xxv.
20. The list of manuscripts is "Catalogue taken of Sr Isaac Newtons M:S:S: Octr: 15th: & 16th: in the Year 1777. By Wm. Godschall Esqr: & the Revd Dr: Horsley," Keynes MS 127A.4. For a fuller description of Horsley's activity at Hurstbourne Park, see Whiteside, *Mathematical Papers*, vol. 1, xxv–xxxviii. The fragment on fluxions is now labeled Add. 3962.6 and resides in the Cambridge University Library.
21. Royal Society (hereafter RS) Minutes of Council, July 9, 1778, vol. 6, fols. 346–50, cited in F. C. Mather, *High Church Prophet: Bishop Samuel Horsley (1733–1806) and the Caroline Tradition in the Later Georgian Church* (Oxford: Oxford University Press, 1992), 43.

## Chapter 4

1. *Buds of Genius, Or, Some Account of the Early Lives of Celebrated Characters: Who Were Remarkable in their Childhood* (London: Darton, Harvey and Darton, 1818), 29–30, emphasis in original.
2. Biot's original French articles was published anonymously as "Newton (Isaac)," in *Biographie universelle, ancienne et moderne*, 83 vols., edited by L. G. Michaud (Paris: Michaud Freres, 1811–53), vol. 31 (1822), 127–94. The English translation by H. Elphinestone in 1829 appeared as Jean-Baptiste Biot, "Life of Sir Isaac Newton," in *Lives of Eminent Persons* (London: Baldwin and Craddock, 1833). Much of the following is indebted to Rebekah Higgitt's comprehensive treatment of Biot's piece and the responses to it in *Recreating Newton: Newtonian Biography and the Making of Nineteenth-Century History of Science* (London: Pickering & Chatto, 2007).
3. Isaac Todhunter, ed., *William Whewell, D.D. Master of Trinity College, Cambridge: An account of his writings; with selections from his literary and scientific correspondence* (1876; Cambridge, UK: Cambridge University Press, 2011), 352.
4. Biot, "Life of Sir Isaac Newton," 26.
5. Biot, "Life of Sir Isaac Newton," 26, quoting from A. de la Pryme, *The Diary of Abraham de la Pryme, the Yorkshire Antiquary*, edited by C. Jackson (London: Publications of the Surtees Society, 1870), vol. 54, 23.
6. Biot, "Life of Sir Isaac Newton," 38, emphasis in original.
7. "Review of Biot's Life of Sir Isaac Newton," *Quarterly Review* 44 (1831): 57.
8. Peter King, *The Life of John Locke, with Extracts from his Correspondence, Journals and Common-Place Books* (London: Henry Colburn, 1829), 224–25.
9. William Whewell, *The Philosophy of the Inductive Sciences Part 1* (Cambridge, UK: John W. Parker J&J Deighton, 1840), cxiii. Whewell referred to the meeting and the coining of the term by "some ingenious gentleman" (himself) who proposed that "by analogy with *artist*, they might form *scientist*," in [W. Whewell], "Review of

Mary Somerville's *On the Connexion of the Physical Sciences,*" *Quarterly Review* 51 (1834): 58–61.

10. David Brewster, *The Life of Sir Isaac Newton,* The Family Library (London: John Murray, 1831), vol. 24, 227.
11. Letter from Brewster to Rigaud, September 15, 1830, Bodleian Library, Oxford (BLO hereafter), MSS Rigaud 60, f.80, cited in Higgitt, *Recreating Newton,* 56.
12. Volume cost from Higgitt, *Recreating Newton,* 209n7.
13. Brewster, *The Life of Sir Isaac Newton,* 234–35.
14. Brewster, *The Life of Sir Isaac Newton,* 329.
15. Benjamin Malkin, "Review of 'The Life of Sir Isaac Newton,' " *Edinburgh Review* 56 (1832): 3.
16. Malkin, "Review," 3.

## Chapter 5

1. Francis Baily, "A short Account of some MSS. Letters (addressed to Mr. Abraham Sharp, relative to the Publication of Mr. Flamsteed's *Historia Coelestis,*) laid on the table, for the inspection of the Members of the Association," in *Report of the 3rd meeting of the British Association for the Advancement of Science, held at Cambridge in July 1833* (London: John Murray, 1834), 462–66.
2. Francis Baily, *Account of the Revd. John Flamsteed, the First Astronomer Royal, to which is added, his British Catalogue of Stars, Corrected and Enlarged* (London: Lords Commissioners of the Admiralty, 1835), xiii.
3. William Whewell, in *Report of the 3rd meeting,* xi–xii.
4. John Herschel, "Memoir of Francis Baily," in Augustus De Morgan, ed., *Journal of a Tour in Unsettled Parts of North America in 1796 & 1797 by the late Francis Baily; with a memoir of the author* (London: Baily Brothers, 1856), 54.
5. Baily, *Account of the Revd. John Flamsteed,* xiv.
6. Baily, *Account of the Revd. John Flamsteed,* xiv–xv.
7. Baily, *Account of the Revd. John Flamsteed,* xxi.
8. Baily, *Account of the Revd. John Flamsteed,* 228.
9. Baily, *Account of the Revd. John Flamsteed,* 323.
10. Baily, *Account of the Revd. John Flamsteed,* 76.
11. Baily, *Account of the Revd. John Flamsteed,* xvi–xvii.
12. Baily, *Account of the Revd. John Flamsteed,* xx.
13. Baily, *Account of the Revd. John Flamsteed,* xix.
14. This insight into Whewell's trajectory comes from John Hodge, "The History of the Earth, Life and Man: Whewell and Palaetiological Science," in *William Whewell: A Composite Portrait,* edited by M. Fisch and S. Schaffer (Oxford: Oxford University Press, 1991), 253–88, 259.
15. William Whewell, *Astronomy and General Physics with reference to Natural Theology* (London: Pickering, 1834), 307. For more on Whewell, see Richard Yeo, *Defining Science: William Whewell, Natural Knowledge and Public Debate in Early Victorian Britain* (Cambridge, UK: Cambridge University Press, 2005).
16. William Whewell, "Remarks on a Note on a Pamphlet entitled 'Newton and Flamsteed' in No. CX of the Quarterly Review" (London: John Parker, 1836), 21–22.
17. Whewell, "Remarks on a Note on a Pamphlet," 28.
18. Cited in Whewell, "Remarks on a Note on a Pamphlet," 6.
19. Thomas Galloway, "Reviews of Three French and English Biographies of Newton," *Foreign Quarterly Review* 12 (1833): 1–27, citation 22.

20. Galloway, "Reviews of Three French and English Biographies of Newton," 17.
21. Letter from Newton Fellowes to Baily, November 24, 1835, Cambridge University Library, Royal Greenwich Observatory Manuscripts (hereafter CUL RGO MSS), Baily Papers, RGO 60/3, cited in Higgitt, *Recreating Newton*, 96.
22. Thomas Galloway, "Life and Observations of Flamsteed—Newton, Halley and Flamsteed," *Edinburgh Review* (1836): 359–97, citation 397.
23. Letter from Brewster to Rigaud, May 21, 1837, MSS Rigaud 60/86, BLO, cited in Higgitt, *Recreating Newton*, 130.
24. Letter from Brewster to Napier, June 14, 1837, Napier Correspondence, Add. MS 34,618, f61, BL, cited in Higgitt, *Recreating Newton*, 131, emphasis in original.
25. Letter from Brewster to Mrs. Liddle, September 12, 1838, David Brewster Papers, Edinburgh Library, cited in Higgitt, *Recreating Newton*, 131.

## Chapter 6

1. Arthur Sherbo, "Heber, Richard (1774–1833)," in *Oxford Dictionary of National Biography* (Oxford: Oxford University Press, 2004).
2. For Dibdin and bibliomania, see Philip Connell, "Bibliomania: Book Collecting, Cultural Politics and the Rise of Literary Heritage in Romantic Britain," *Representations* 71 (2000): 24–47.
3. H. R. Luard, "Thomas Dibdin," in *Dictionary of National Biography* (London: Smith, Elder, 1888).
4. Philippa Levine, *The Amateur and the Professional: Antiquarians, Historians and Archaeologists in Victorian England, 1838–1886* (Cambridge, UK: Cambridge University Press, 1986), 41.
5. Edward Edwards, *Libraries and the Founders of Libraries* (London: Trubner, 1864), 446.
6. Edwards, *Libraries and the Founders of Libraries*, 446.
7. Edwards, *Libraries and the Founders of Libraries*, 406.
8. Thomas Frognall Dibdin, *Bibliophobia: Remarks on the Present Languid and Depressed State of Literature and the Book Trade. In a Letter Addressed to the Author of the Bibliomania by Mercurious Rusticus* (London: Henry Bohn, 1832), 10–11.
9. A. N. L. Munby, *The Cult of the Autograph Letter in England* (London: Athlone Press, 1962), 7.
10. Upcott to Turner, June 2, 1820, cited in Munby, *Cult*, 32; Janet Ing Freeman, "Upcott, William (1779–1845)," in *Oxford Dictionary of National Biography* (Oxford: Oxford University Press, 2004).
11. Review, *London and Edinburgh Philosophical Review*, third series, 13 (1838): 221.
12. Theodore Hornberger, "Halliwell-Phillipps and the History of Science," *Huntington Library Quarterly* 12 (1949): 391–99, citation 392.
13. Arthur Freeman and Janet Ing Freeman, "Phillipps, James Orchard Halliwell- (1820–1889)," in *Oxford Dictionary of National Biography* (Oxford: Oxford University Press, 2004).
14. For the material on Halliwell-Phillipps and the Historical Society of Science, I have relied on A. N. L. Munby, *The History and Bibliography of Science in England; The First Phase, 1833–1845* (Berkeley: University of California Press, 1968).
15. Review, *London, Edinburgh and Dublin philosophical magazine and journal of science*, third series, 18, (1841): 412–13.
16. [Augustus De Morgan], "Review of 'A Collection of Letters, Illustrative of the Progress of Science in England, from the Reign of Elizabeth to that of Charles II.' Edited by J. O. Halliwell," *Athenaeum* 2 (1841): 588–89, citation 588.
17. De Morgan, "Review," 589.

18. Stephen Peter Rigaud, *Historical Essay on the First Publication of Sir Isaac Newton's Principia* (Oxford: Oxford University Press, 1838), vii.
19. Stephen Peter Rigaud, *Correspondence of Scientific Men of the Seventeenth Century, including Letters of Barrow, Flamsteed, Wallis and Newton* (Oxford: Oxford University Press, 1841), v.
20. Rigaud, *Correspondence of Scientific Men of the Seventeenth Century*, viii.
21. Joseph Edleston, *Correspondence of Sir Isaac Newton and Professor Cotes* (London: John Parker, 1850), lxxxvi.
22. Edleston, *Correspondence of Sir Isaac Newton and Professor Cotes*, lxii.
23. Augustus De Morgan, *Arithmetical Books from the Invention of Printing to the Present Time* (London: Taylor and Walton, 1847), ii.
24. Sophia De Morgan, *Memoir of Augustus De Morgan* (London: Longmans, Green, 1882), 256.
25. Augustus De Morgan, *Essays on the Life and Work of Newton* (Chicago: Open Court, 1914), 37. This edition is a reprint of De Morgan's original publication "Newton," in *Cabinet Portrait Gallery of British Worthies* 11 (1846): 78–117.
26. De Morgan, *Memoir of Augustus De Morgan*, citing Augustus De Morgan, 258.
27. De Morgan, *Essays on the Life and Work of Newton*, 38.
28. Baden Powell, "Review of recent books on Newton," *Edinburgh Review* (1856): 499–534, citation 500.
29. David Brewster, *Memoirs of the Life, Writings, and Discoveries of Isaac Newton* (Edinburgh: Thomas Constable, 1855), x.
30. Brewster, *Memoirs of the Life, Writings, and Discoveries of Isaac Newton*, xii.
31. Brewster, *Memoirs of the Life, Writings, and Discoveries of Isaac Newton*, xv.
32. Powell, "Review of recent books on Newton," 511.
33. Brewster, *Memoirs of the Life, Writings, and Discoveries of Isaac Newton*, 374.
34. Newton to Locke, February 7, 1691, cited in Brewster, *Memoirs of the Life, Writings, and Discoveries of Isaac Newton*, 318.
35. Brewster, *Memoirs of the Life, Writings, and Discoveries of Isaac Newton*, 332.
36. Brewster, *Memoirs of the Life, Writings, and Discoveries of Isaac Newton*, 337.
37. Brewster, *Memoirs of the Life, Writings, and Discoveries of Isaac Newton*, 337n1.
38. Brewster, *Memoirs of the Life, Writings, and Discoveries of Isaac Newton*, 340.
39. Margaret Maria Gordon, *The Home Life of Sir David Brewster*, 2nd edition (Edinburgh: Edmonston and Douglas, 1870), 262.
40. Gordon, *The Home Life of Sir David Brewster*, 262.
41. Letter from Brewster to Brougham, July 11, 1854, Brougham Papers, 26,679, University College London, cited in Higgitt, *Recreating Newton*, 144.
42. Brewster, *Memoirs of the Life, Writings, and Discoveries of Isaac Newton*, 355.
43. Augustus De Morgan, "Review of Sir David Brewster's Life of Newton," *North British Review* (1855): 307–38, citation 336.
44. De Morgan, "Review of Sir David Brewster's Life of Newton," 333.
45. De Morgan, "Review of Sir David Brewster's Life of Newton," 309.
46. Gordon, *The Home Life of Sir David Brewster*, 260–61.
47. De Morgan, *Essays on the Life and Work of Newton*, 51.
48. De Morgan, *Essays on the Life and Work of Newton*, 63.
49. Augustus De Morgan, *Newton: His Friend: and His Niece* (London: Dawsons, 1885).
50. De Morgan, *Newton*, vi.
51. Gordon, *The Home Life of Sir David Brewster*, 410. The famous saying attributed to Newton is not to be found anywhere in the Newtonian corpus and may be a bit of Brewsterian apocrypha.

## Chapter 7

1. *A Catalogue of the Portsmouth Collection of Books and Papers, written by or belonging to Sir Isaac Newton, The scientific portion of which has been presented by the Earl of Portsmouth to the University of Cambridge, drawn up by the Syndicate appointed the 6th November 1872* (Cambridge, UK: Cambridge University Press, 1888), ix.
2. Evidently John Conduitt's original bond promising any proceeds of the papers for Newton's heirs had by now been forgotten.
3. John Couch Adams diary, July 30, 1872, St. John's Library/Adams/Box 21/11.
4. James Glaisher, "Memoir of the Life of John Couch Adams," in *The Scientific Papers of John Couch Adams*, edited by W. G. Adams (Cambridge, UK: Cambridge University Press, 1896), xlvi.
5. Lord Portsmouth to Vice-Chancellor Taylor, August 2, 1872, CUL Add. MS 2588.495.
6. Lord Portsmouth to Vice Chancellor Taylor, August 2, 1872, CUL Add. MS 2588.495.
7. Lady Portsmouth to Vice Chancellor Taylor, August 6, 1872, CUL Add. MS 2588.496.
8. *First Report of the Royal Commission on Historical Manuscripts*, Parliamentary Papers (hereafter PP) 1870 [C.55], 3.
9. *Second Report of the Royal Commission on Historical Manuscripts*, PP 1871 [C.441], xxii.
10. *Third Report of the Royal Commission on Historical Manuscripts*, PP 1872 [C.673], xxvi–xxvii.
11. *First Report*, ix.
12. *First Report*, x.
13. *First Report*, xi.
14. *Second Report*, x.
15. *Third Report*, xxvi–xxvii.
16. Lord Portsmouth to Vice-Chancellor Taylor, July 23, 1872, CUL Add. MS 2588/494.
17. Much of the following discussion on the Mathematical Tripos in nineteenth-century Cambridge is indebted to Andrew Warwick, *Masters of Theory: Cambridge and the Rise of Mathematical Physics* (Chicago: University of Chicago Press, 2003). Both Rayleigh and Maxwell used student exam papers to draft their own scientific papers Warwick, *Masters of Theory* (19).
18. William Whewell, ed., *Principia, Book 1, Sections I. II. And III. In the original Latin with explanatory notes and references* (London: John Parker, 1846), iii.
19. Warwick, *Masters of Theory*, 205.
20. On coaching, see Warwick, *Masters of Theory*, 227–85.
21. Cited in Warwick, *Masters of Theory*, 196.
22. For more on Stokes's contribution to viscous theory, see Olivier Darrigol, "Between Hydrodynamics and Elasticity Theory: The First Five Births of the Navier-Stokes Equation," *Archive of the Exact Sciences* 56 (2002): 139–45.
23. Cited in Morton Grosser, *The Discovery of Neptune* (New York: Dover, 1979), 74.
24. Anonymous undergraduate named "Peregrine" in the *Queen*, November 11, 1893, cited in William Sheehan and Steven Thurber, "John Couch Adams's Asperger Syndrome and the British Non-Discovery of Neptune," *Notes and Records of the Royal Society of London* 61 (2007): 285–99, citation 292.
25. Glaisher, "Memoir of the Life of John Couch Adams," xliii, n1.
26. Glaisher, "Memoir of the Life of John Couch Adams," xlv.
27. Sheehan and Thurber, "John Couch Adams's Asperger Syndrome," 292.
28. Galileo had in fact observed the planet in 1612 but had taken it to be a fixed star.

29. Airy to Sedgwick, December 4, 1846, cited in Sheehan and Thurber, "John Couch Adams's Asperger Syndrome," 294.
30. For more on the complicated story of Neptune's discovery, see N. R. Hanson, "Leverrier: The Zenith and Nadir of Newtonian Mechanics," *Isis* 53 (1962): 359–78.
31. Challis to Airy, December 19, 1846, cited in Sheehan and Thurber, "John Couch Adams's Asperger Syndrome," 294.
32. Glaisher, "Memoir of the Life of John Couch Adams," xxxii.
33. Glaisher, "Memoir of the Life of John Couch Adams," xxxi.
34. Glaisher, "Memoir of the Life of John Couch Adams," xxxii.
35. John Conduitt, "Notes on Newton's character," Keynes Ms. 130.07, UK6v.
36. J. C. Adams, "On the Secular Variation of the Moon's Mean Motion," *Philosophical Transactions of the Royal Society* 143 (1853): 397–406.
37. Airy to G. H. Richards, September 13, 1872, CUL, RGO 6 150/158.
38. George Gabriel Stokes, *Memoirs and Scientific Correspondence of the Late Sir George Gabriel Stokes*, edited by Joseph Larmor (Cambridge, UK: Cambridge University Press, 1907), vol. 1, 72.
39. Stokes to Mary Stokes, January 17, 1857, in Stokes, *Memoirs and Scientific Correspondence*, 52.
40. Stokes to Mary Stokes, April 1, 1857, in Stokes, *Memoirs and Scientific Correspondence*, 63.
41. James Glaisher, "Obituary of John Couch Adams," *Observatory* 15 (1892): 173–89, citation 173.
42. Stokes, *Memoirs and Scientific Correspondence*, 274.
43. Kelvin to Stokes, November 30, 1884; Stokes to Kelvin, December 6, 1884, in *The Correspondence between Sir George Gabriel Stokes and Sir William Thomson, Baron Kelvin of Largs*, edited by David Wilson (Cambridge, UK: Cambridge University Press, 2011), 572–74.
44. [Chemicus], "The late Sir G. G. Stokes," *Nature* 67 (February 19, 1903): 367.
45. Stokes, *Memoirs and Scientific Correspondence*, 34–35.
46. Stokes, *Memoirs and Scientific Correspondence*, 25.
47. Glaisher, "Memoir of the Life of John Couch Adams," xliii.
48. Glaisher, "Memoir of the Life of John Couch Adams," xlv.

## Chapter 8

1. J. Willis Clark, *Old Friends at Cambridge and Elsewhere* (London: Macmillan, 1900), 338.
2. *Eighth report of the Royal Commission on Historical Manuscripts*, PP 1881 [C. 3040], xi–xii.
3. Letter from Taylor to Adams, March 8, 1887, St. John's College, Cambridge, Papers of John Couch Adams (hereafter JCA), 27/1/8.1. By kind permission of the Master and Fellows of St. John's College.
4. Letter from Taylor to Adams, March 11, 1887, JCA 27/1/8.2.
5. Letter from Taylor to Adams, March 16, 1887, JCA 27/1/8.3.
6. Letter from Taylor to Adams, July 6, 1887, JCA 27/1/8.4.
7. Preface to *A Catalogue of the Portsmouth Collection of Books and Papers written by or belonging to Sir Isaac Newton, the Scientific Part of which has been presented by the Earl of Portsmouth to the University of Cambridge, drawn up by the Syndicate appointed the 6th November, 1872* (Cambridge, UK: Cambridge University Press, 1888), ix.

8. For more on Newton's method, see Niccolo Guiciardini, *Isaac Newton on Mathematical Certainty and Method* (Cambridge, MA: MIT Press, 2009).
9. Preface, xii.
10. J. C. Adams, "On the Secular Variation of the Eccentricity and Inclination of the Moon's Orbit," *Monthly Notices of the Royal Astronomical Society* 19 (1859): 206–8.
11. Preface, xv.
12. Preface, ix–x.
13. Preface, xix–xx.
14. George Howard Darwin, "Inaugural Lecture on Election to the Plumian Professorship," in *Scientific Papers by Sir George Howard Darwin*, edited by F. J. M. Stratton and J. Jackson (Cambridge, UK: Cambridge University Press, 1916), vol. 5, 5–6.
15. Glaisher, "Memoir of the Life of John Couch Adams," xliii.
16. Adams's scientific papers were published as *The Scientific Papers of John Couch Adams*, 2 vols., edited by W. G. Adams and R. A. Sampson (London: Cambridge University Press, 1896–1900), with a memoir by J. W. L. Glaisher.
17. Stokes, *Memoirs and Scientific Correspondence*, v.
18. Lord Rayleigh, "Obituary Notice," in *Mathematical and Physical Papers by the Late Sir George Gabriel Stokes* (Cambridge, UK: Cambridge University Press, 1903), vol. 5, xxiv.
19. Stokes, *Memoirs and Scientific Correspondence*, v.

## Chapter 9

1. The lone exception was a couple of works based on the papers published in the 1890s by William Rouse Ball, a Trinity College tutor and amateur historian of mathematics: William Rouse Ball, "On Newton's Classification of Cubic Curves," *Proceedings of the London Mathematical Society* 22 (1891): 104–43; William Rouse Ball, "A Newtonian Fragment Relating to Centripetal Forces," *Proceedings of the London Mathematical Society* 23 (1892): 226–31; William Rouse Ball, *An Essay on Newton's "Principia"* (London, 1893).
2. Biographical information on Wallop is from Gerard Wallop, *A Knot of Roots: An Autobiography* (New York: Dutton, 1965).
3. *Times*, January 21, 1936.
4. Sanderson to Lymington, February 26, 1936, Hampshire Record Office, 15M84/F390, cited in Dan Stone, "The English Mistery, the BUF, and the Dilemmas of British Fascism," *Journal of Modern History* 75 (2003): 336–58, citation 345.
5. *Times*, June 23, 1936.
6. *Daily Telegraph*, June 22, 1936.
7. *Times*, April 20, 1935.
8. A. N. L. Munby, "The Library," in *The Destruction of the Country House*, edited by Roy Strong (London: Thames and Hudson, 1974), 106–10, citation 106.
9. Munby, "The Library," 107.
10. Munby, "The Library," 107.
11. *Times*, August 14, 1848.
12. Giles Worsley, *England's Lost Houses: From the Archives of Country Life* (London: Aurum Press, 2011), 11.
13. Peter Reid, "The Decline and Fall of the British Country House Library," *Libraries and Culture* 36 (2001): 345–66, 355.
14. Munby, "The Library," 107.
15. Reid, "The Decline and Fall of the British Country House Library," 352.

16. A. S. W. Rosenbach to Sir R. Leicester Harmsworth, November 3, 1933, Rosenbach Company Archives, I:82-O1, cited in Leslie Morris, *Rosenbach Redux: Further Book Adventures in England and Ireland* (Philadelphia: Rosenbach Museum and Library, 1989), 69.
17. Roger Babson, *Actions and Reactions* (New York: Harper, 1949), 10.
18. Babson, *Actions and Reactions,* 17.
19. Roger Babson, *Cheer Up! Better Times Ahead* (New York: Fleming H. Revell, 1932), 10.
20. Babson, *Actions and Reactions,* 136.
21. Grace K. Babson Collection of Newtoniana, Huntington Library, San Marino, CA, Babson Newton 17—Curatorial Correspondence, Misc Correspondence, 1925–60, Folder labeled "1925–1930 Correspondence about the Newton collection."
22. Babson, *Actions and Reactions,* 340.
23. Babson, *Actions and Reactions,* 342.
24. Roger Babson, "Gravity—Our Enemy Number One," included as appendix to Harry Collins, *Gravity's Shadow: The Search for Gravitational Waves* (Chicago: University of Chicago Press, 2004), 828–31, citation 828.
25. Babson, "Gravity—Our Enemy Number One," 828.
26. "Summer Conferences on Investments and Gravity New Boston, New Hampshire, August 27–31, 1960 No Admission Charge," flier, accessed September 23, 2013, at http://www.newbostonhistoricalsociety.com/gravity.html.
27. John Gribbin and Michael White, *Stephen Hawking, a Life in Science* (Washington, DC: Joseph Henry Press, 2002), 150–51. Gribbin, a former physicist turned science writer, himself won the prize in 1970.

## Chapter 10

1. Ernst Weil, "Milestones of Civilization," in *Talks on Book-Collecting: Delivered under the Authority of the Antiquarian Booksellers Association,* edited by P. H. Muir (London: Cassell, 1952), 84.
2. John Carter reports that "the preponderance of professional over private bidding gave a stability to prices which the dealers had a real and personal interest in maintaining, in bad times as in good." John Carter, *Taste and Technique in Book-Collecting: A Study of Recent Developments in Great Britain and the United States* (Cambridge, UK: Cambridge University Press, 1949), 128.
3. Interview with Robin Waterfield, in Sheila Markham, *A Book of Booksellers: Conversations with the Antiquarian Book Trade, 1991–2003* (London: Oak Knoll Press, 2007), 243.
4. For a detailed analysis of the sale at Ruxley Lodge, see Arthur Freeman and Janet Ing Freeman, *Anatomy of an Auction: Rare Books at Ruxley Lodge, 1919* (London: Book Collector, 1990).
5. H[einrich] Z[eitlinger], *Bibliotheca Chemico-Mathematica* (London: Henry Sotheran, 1921).
6. Zeitlinger, *Bibliotheca,* vol. 1, v.
7. Zeitlinger, *Bibliotheca,* vol. 1, v.
8. John L. Heilbron, "Interview with E. N. da C. Andrade," December 18, 1962, accessed February 23, 2013, at http://www.aip.org/history/ohilist/4488.html.
9. Heinrich Zeitlinger, *Bibliotheca,* Third Supplement (London: H. Sotheran, 1952), preface.
10. Zeitlinger, *Bibliotheca,* Third Supplement, preface.
11. John Hill Burton, *The Book Hunter etc.* (London: Blackwoods, 1862), 209.

12. Carter, *Taste and Technique in Book-Collecting*, 6.
13. E. Weil, "E. P. Goldschmidt: Bookseller and Scholar," *Journal of the History of Medicine and Allied Sciences* 2 (1954): 224–32, citation 230.
14. "Obituary of E. P. Goldschmidt," *Antiquarian Bookman*, March 6, 1954, 638.
15. E. P. Goldschmidt, "The Period before Printing," in *Talks on Book-Collecting*, edited by P. H. Muir (London: Cassell, 1952), 25–38, citation 25.
16. Herbert McLean Evans, ed., *Exhibition of First Editions of Epochal Achievements in the History of Science* (Berkeley: University of California Press, 1934).
17. Carter, *Taste and Technique in Book-Collecting*, 69.
18. Carter, *Taste and Technique in Book-Collecting*, 55.
19. Weil, "Milestones of Civilization," 84.

### Chapter 11

1. [J. C. Taylor], *Catalogue of the Newton Papers Sold by Order of the Viscount Lymington* (London: Sotheby's, 1936). See Peter Spargo, "Sotheby's, Keynes and Yahuda—The 1936 Sale of Newton's Manuscripts," in *The Investigation of Difficult Things*, edited by P. Harman and A. Shapiro (Cambridge, UK: Cambridge University Press, 1992), 115–34.
2. *Times*, June 23, 1936.
3. A. N. L. Munby, "Book Collecting in the 1930s," in Munby, *Essays and Papers* (London: Scolar Press, 1977), 218.
4. Munby, "Book Collecting," 218.
5. Robert Skidelsky, *John Maynard Keynes* (New York: Viking, 1986), vol. 1, 116.
6. Keynes to B. W. Swithinbank, November 13, 1902, cited in Roy Harrod, *The Life of John Maynard Keynes* (London: Macmillan, 1951), 68.
7. Munby, "Book Collecting," 218.
8. A. N. L. Munby, "Keynes as a Book Collector," in *Essays on John Maynard Keynes*, edited by Milo Keynes (Cambridge, UK: Cambridge University Press, 1975), 290–98.
9. Robert Skidelsky, *John Maynard Keynes, 1883–1946: Economist, Philosopher, Statesman* (New York: Pan, 2004), 466.
10. John Maynard Keynes, "On Reading Books," in *A Bloomsbury Group Reader*, edited by S. P. Rosenbaum (Cambridge, MA: Blackwell, 1993), 286–91.
11. Wallop to Keynes, September 8, 1936, King's College Library Archives (hereafter KCL), JMK-67-PP-60-f19-20.
12. Invoices are at KCL, JMK-PP-58-88/89.
13. Maggs Bros.' dominance at the sale makes sense. They were often granted an extended line of credit with Sotheby's, so it was relatively painless for them to buy material without a ready buyer, as a way of building their stock. Personal email correspondence with Robert Harding of Maggs Bros., July 17, 2012.
14. Handwritten notes, KCL, JMK-67-PP-60-67.
15. *Times*, May 1, 1936.
16. *Times Literary Supplement*, July 18, 1936, 604.
17. *Times*, July 27, 1936.
18. Invoice from Edwards to Keynes, August 18, 1936, KCL, JMK-67-PP-58/2; Keynes to Edwards, August 25, 1936, KCL, JMK-67-PP-58/3.
19. Keynes to Fabius, August 12, 1936, KCL, JMK-67-PP-58-7.
20. Keynes to Ernest Maggs, August 3, 1936, KCL, JMK-67-PP-58-39.
21. Keynes to Ernest Maggs, August 12, 1936, KCL, JMK-67-PP-58-46.

22. Keynes to Goldschmidt, August 3, 1936, KCL, JMK-67-PP-58-15; Keynes to Heffer, August 3, 1936, KCL, JMK-67-58-19.
23. Heffer to Keynes, August 4, 1936, KCL, JMK-67-PP-58-20.
24. Keynes to Heffer, August 6, 1936, KCL, JMK-67-PP-28-20.
25. Wallop to Keynes, September 8, 1937, KCL, JMK-67-PP-58-19.
26. Keynes to Wallop, September 9, 1936, KCL, JMK-67-PP-58-21.
27. Keynes to Fabius, September 17, 1936, KCL, JMK-67-PP-58-12; for quote, Keynes to Wallop, September 9, 1936.
28. Keynes to Maggs, August 19, 1936, KCL, JMK-67-PP-58-56.
29. Keynes to Wells, August 3, 1936, KCL, JMK-67-PP-58-165; Wells to Keynes, August 18, 1936, KCL, JMK-67-PP-58-166.
30. Keynes to Wells, August 25, 1936, KCL, JMK-67-PP-58-168.
31. Keynes to Wells, September 8, 1936, KCL, JMK-67-PP-58-170.
32. Maggs to Keynes, September 7, 1936, KCL, JMK-67-PP-58-176.
33. Keynes to Yahuda, September 9, 1936, KCL, JMK-67-PP-58-177.
34. Yahuda to Keynes, September 15, 1936, KCL, JMK-67-PP-58-179.
35. Keynes to Yahuda, November 4, 1936, KCL, JMK-67-PP-58-181.
36. Keynes to Yahuda, April 3, 1938, KCL, JMK-67-PP-58-186.
37. Yahuda to Keynes, April 24, 1938, KCL, JMK-67-PP-58-188.
38. Yahuda to Keynes, July 24, 1938, KCL, JMK-67-PP-58-191.
39. Keynes to L. F. Gilbert, September 28, 1937, KCL, JMK-67-PP-60-f13.
40. L. F. Gilbert to Keynes, September 29, 1937, KCL, JMK-67-PP-60-f14.
41. Gilbert to Keynes, November 9, 1937, KCL, JMK 67-PP-60-f17.
42. Keynes loaned nine letters from Halley to Newton relating to the publication of the first edition of the *Principia,* as well as an alchemical MS for the exhibit that accompanied the celebration.
43. See John Russell to Keynes, December 4, 1942, KCL, JMK-67-PP-60-179; Broad to Keynes, December 7, 1942, KCL, JMK-67-PP-60-185.
44. Keynes to Broad, December 19, 1942, KCL, JMK-67-PP-60-186.
45. Keynes to Henry Dale, December 11, 1942, KCL, JMK-67-PP-60-184.
46. John Maynard Keynes, "Newton the Man," in *The Collected Writings of John Maynard Keynes,* edited by Donald Moggridge and Elizabeth Johnson (Cambridge, UK: Cambridge University Press, 1972), vol. 10, 363–74.
47. Keynes, "Newton the Man."
48. John Maynard Keynes, "Einstein," in *Essays in Biography* (London: Palgrave Macmillan, 2010).

## Chapter 12

1. Letter from Yahuda to Ethel Yahuda, July 23, 1936, National Library of Israel Archives (hereafter NLI), MS. VAR. Yah 38.3164: Ethel-AS Yahuda Correspondence, 1936.
2. Letter from Yahuda to Ethel Yahuda, July 28, 1936, NLI MS. VAR. Yah 38.3164: Ethel-AS Yahuda Correspondence, 1936.
3. Letter from Yahuda to Ethel Yahuda, July 31, 1936, NLI MS. VAR. Yah 38.3164: Ethel-AS Yahuda Correspondence, 1936.
4. Letter from Yahuda to Ethel Yahuda, August 27, 1936, NLI MS. VAR. Yah 38.3164: Ethel-AS Yahuda Correspondence, 1936.
5. Letter from Yahuda to Ethel Yahuda, August 30, 1936, NLI MS. VAR. Yah 38.3164: Ethel-AS Yahuda Correspondence, 1936.

6. Letter from Yahuda to Ethel Yahuda, September 25, 1936, NLI MS. VAR. Yah 38.3164: Ethel-AS Yahuda Correspondence, 1936.
7. On Yahuda, see Reeva Spector Simon, Michael Lasker, and Sara Reguer, eds., *The Jews of the Middle East and Africa in Modern Times* (New York: Columbia University Press, 2003), 86–87; Martin Plessner, "Abraham Shalom Yahuda," in *Encyclopedia Judaica* (New York: Macmillan, 1972), vol. 2, 272; Michael Fishbane and Judith Weschler, eds., *The Memoirs of Nahum N. Glatzer*, vol. 6 of *Jewish Perspectives* (Cincinnati: Hebrew Union College Press, 1997), 107.
8. Fishbane and Weschler, *Memoirs*, 105.
9. Fishbane and Weschler, *Memoirs*, 107.
10. Fishbane and Weschler, *Memoirs*, 108.
11. NLI, Yah. Ms. Var. 1/Newton Papers 43/3, 7 N2–N15.
12. NLI, Yah. Ms. Var. 1/Newton Papers 43/3, 7 N2–N15.
13. NLI, Yah. Ms. Var. 1/Newton Papers 43/3, 7 N2–N15.
14. Abraham Yahuda, *The Accuracy of the Bible: The Stories of Joseph, the Exodus and Genesis Confirmed and Illustrated by Egyptian Monuments and Language* (New York: E. P. Dutton, 1935), viii.
15. Yahuda, *The Accuracy of the Bible*, xxviii.
16. Yahuda, *The Accuracy of the Bible*, xxviii.
17. Online catalogue description at NLI.
18. Yahuda, *The Accuracy of the Bible*, ix.
19. Yahuda, *The Accuracy of the Bible*, xxi.
20. Yahuda, *The Accuracy of the Bible*, xxi.
21. Yahuda, Accuracy *The Accuracy of the Bible*, xxii.
22. From "Excerpts from an Address on 'The Accuracy of the Bible in the Light of Egyptian Antiquity,' delivered by Professor Abraham S. E. Yahuda of London at a Reception in his honor on 27 January 1941, at the Community House of Temple Emanu-El, 1 East 65 Street, New York, NY, Professor Albert Einstein Chairman Reception Committee," Princeton University Library, Department of Rare Books and Special Collections, CO 627, Box 3/Treasure Room Related Material.
23. Fishbane and Weschler, *Memoirs*, 108.
24. NLI, Yah MS Var 1/Newton Papers 43/3.7 N2–N15.
25. NLI, Yah MS Var 1/Newton Papers 43/3.7 N2–N15.
26. Yahuda MS Varia 1, Newton Papers 42, Dept of MSS and Archives, JNUL; see also Albert Einstein Archives, The Hebrew University of Jerusalem, Israel, 69–54 and 69–55.
27. I. Bernard Cohen, "An Interview with Einstein," *Scientific American* 193 (1955): 68–73, citation 72.
28. Fishbane and Weschler, *Memoirs*, p 109.
29. Rudolph Mach, *Catalogue of Arabic Manuscripts (Yahuda section) in the Garrett Collection, Princeton University Library* (Princeton: Princeton University Press, 1977); Efraim Wust, "A Catalogue of the Arabic Manuscripts in the A. S. Yahuda Collection, Jewish National and University Library, Jerusalem, Limited Preliminary Edition," Jerusalem, June 1997.
30. *New York Times*, August 21, 1951.
31. Unmarked yellow folder relating to Yahuda Manuscript Collection, National Library of Israel, Document 1 [top left in pencil "02-6710403"; top right in pencil "David Castillejo 1969"]. I am grateful to Micah Anshan for locating and sharing this document with me.
32. For details of the case, see *Hebrew Assn. v. Nye*, Supreme Court of Connecticut.

## Chapter 13

1. J.L.E. Dreyer, "On the Desirability of Publishing a New Edition of Isaac Newton's Collected Works," *Monthly Notices of the Royal Astronomical Society* 84 (1934): 298–304."
2. R. A. Sampson, "On Editing Newton," *Monthly Notices of the Royal Astronomical Society* 84 (1934): 378–83.
3. George Sarton, "Introduction to the History and Philosophy of Science (Preliminary Note)," *Isis* 4 (1921): 23–31, citation 25.
4. George Sarton, "An Institute for the History of Science and Civilization (Third Article)," *Isis* 28 (1938): 7–17, citation 17.
5. A. Koyré, "The Significance of the Newtonian Synthesis," *Journal of General Education* 4 (1950): 257.
6. The first person to receive a PhD in the history of science in America was Aydin Sayili, a Turkish citizen, who completed his dissertation in 1942 under the supervision of George Sarton. See G. A. Russell, "Aydin Sayili, 1913–1993," *Isis* 87 (1996): 672–75.
7. Joseph W. Dauben, Mary Louise Gleason, and George E. Smith, "Seven Decades of History of Science: I. Bernard Cohen (1914–2003), Second Editor of Isis," *Isis* 100 (2009): 4–35.
8. H. W. Turnball et al., eds., *The Correspondence of Isaac Newton* (Cambridge, UK: Cambridge University Press, 1959–77); D. T. Whiteside, ed., *The Mathematical Papers of Isaac Newton*, 8 vols. (Cambridge, UK: Cambridge University Press, 1967–81). Volumes 2–7 were produced with the assistance in publication of M. Hoskin and A. Prag, volume 8 with that of Prag.
9. E. N. da C. Andrade, Introduction to volume 1 of *Correspondence of Isaac Newton* (Cambridge, UK: Cambridge University Press, 1959), xvii.
10. Andrade, introduction, xviii.
11. I. B. Cohen, Preface to *Isaac Newton's Papers and Letters on Natural Philosophy* (Cambridge, MA: Harvard University Press, 1978).
12. I. B. Cohen, *Introduction to Newton's "Principia"* (Cambridge, MA: Harvard University Press, 1971), 21–22.
13. A. Koyré and I. B. Cohen, eds., with the assistance of Anne Whitman, *Isaac Newton's Philosophiae Naturalis Principia Mathematica, the Third Edition with Variant Readings* (Cambridge, MA: Harvard University Press, 1972).
14. Cohen, *Introduction*, 31–32.
15. Cohen, *Introduction*, xiii.
16. See Michael Hunter, ed., *Archives of the Scientific Revolution: The Formation and Exchange of Ideas in Seventeenth Century Europe* (Woodbridge, UK: Boydell Press, 1998).
17. Details are from a letter to the author from David Castillejo, April 5, 2012.
18. "Notes of experiments in chemistry and alchemy," MS Add. 3973, and "Laboratory notebook," MS Add. 3975, CUL.
19. See MS Add. 3975 f83, CUL, for Newton's reference to the net.
20. Richard S. Westfall, "The Changing World of the Newtonian Industry," *Journal of the History of Ideas* 37 (1976): 175–84, citation 175n2. The citation is an addition to Proposition VII, Problem 11 in the *Principia*, which states, "Let a body orbit in the circumference of a circle; there is required the law of centripetal force tending to any given point whatever." For a discussion of folios 181–87, see D. T. Whiteside, ed., *The Mathematical Papers of Isaac Newton* (Cambridge, UK: Cambridge University Press, 1981), vol. 6, 542–43, 546–57.

21. [Augustus De Morgan], "Review of 'A Collection of Letters, Illustrative of the Progress of Science in England, from the Reign of Elizabeth to that of Charles II.' Edited by J. O. Halliwell," *Athenaeum* 2 (1841): 588–89, citation 589.
22. Whiteside's dissertation was published as "Patterns of Mathematical Thought in the Later Seventeenth Century," *Archive for the History of Exact Sciences* 1 (1961): 179–388.
23. Biographical information on Whiteside is from Niccolo Guicciardini, "In Memoriam, Derek Thomas Whiteside (1932–2008)," *Historia Mathematica* 36 (2009): 4–9.
24. Richard Westfall commented, "Tom [Whiteside] has set a permanent mark on Newtonian scholarship. It is impossible to imagine that anyone will ever feel the need to do the edition anew. As long as Newton continues to be studied—and who can foresee the day when he will not?—the Whiteside edition of his Mathematical Papers will remain a basic datum." Richard S. Westfall, "Award of the 1977 Sarton Medal to D. T. Whiteside," *Notes and Records of the Royal Society* 69 (1978): 86–87, citation 87.
25. Westfall, "The Changing World of the Newtonian Industry," 176.
26. Such as, for example, Niccolo Guicciardini's *Reading the Principia: The Debate on Newton's Mathematical Methods for Natural Philosophy from 1687 to 1736* (Cambridge, UK: Cambridge University Press, 1999).
27. Westfall, "The Changing World of the Newtonian Industry," 176.
28. Frances Yates, *Giordano Bruno and the Hermetic Tradition* (London: Routledge & Kegan, Paul, 1964).
29. J. E. McGuire and P. M. Rattansi, "Newton and the 'Pipes of Pan,'" *Notes and Records of the Royal Society* 21 (1966): 108–43.
30. J. E. McGuire, "Force, Active Principles, and Newton's Invisible Realm," *Ambix* 15 (1968): 154–208.
31. D. C. Kubrin, "Newton and the Cyclical Cosmos: Providence and the Mechanical Philosophy," *Journal of the History of Ideas* 28 (1967): 325–46.
32. B. J. T. Dobbs, *The Foundations of Newton's Alchemy or "The Hunting of the Greene Lyon"* (Cambridge, UK: Cambridge University Press, 1975).
33. F. E. Manuel, *Isaac Newton, Historian* (Cambridge, MA: Belknap Press of Harvard University Press, 1963); F. E. Manuel, *A Portrait of Isaac Newton* (Cambridge, MA: Belknap Press of Harvard University Press, 1968).
34. F. E. Manuel, *The Religion of Isaac Newton* (Oxford: Oxford University Press, 1974).
35. R. S. Westfall, *Never at Rest: A Biography of Isaac Newton* (Cambridge, UK: Cambridge University Press, 1980).
36. John Harrison, *The Library of Isaac Newton* (Cambridge, UK: Cambridge University Press, 1978).
37. A. R. Hall, "Newton's Revolution," *British Journal for the Philosophy of Science* 33 (1982): 305–15, citation 307.
38. R.4.48c Trinity College Library, Cambridge.
39. MS Add. 3975 f23, CUL.
40. Michael Mahoney, "'On Differential Calculuses,' *The Mathematical Papers of Isaac Newton, Vol VIII, 1697–1722*," *Isis* 75 (1984): 366–72, citation 371.
41. For a discussion of Newton's solitude, and lack thereof, see Rob Iliffe, "'Is He Like Other Men?' The Meaning of the *Principia mathematica* and the Author as Idol," in *Culture and Society in the Stewart Restoration: Literature, Drama, History*, edited by Gerald MacLean (Cambridge, UK: Cambridge University Press, 1995), 159–76.

42. Humphrey Newton, "Two Letters from Humphrey Newton to John Conduitt," Keynes Ms. 135, King's College, Cambridge; John Conduitt, "Notes on Newton's Character," Keynes Ms. 130.07, King's College, Cambridge.
43. Isaac Newton, "Extract from Treatise on Revelation," Yah MS 9.2 f 140, NLI.

## Chapter 14

1. Manuel, *The Religion of Isaac Newton*, 103.
2. Betty Jo Dobbs, *The Janus Face of Genius: The Role of Alchemy in Newton's Thought* (Cambridge, UK: Cambridge University Press, 1991) 6.
3. Penelope M. Gouk, "Review of David Castillejo 'The Expanding Force in Newton's Cosmos, as Shown in His Unpublished Papers,'" *British Journal for the History of Science* 17 (1984): 112–13, citation 113.
4. David Castillejo, *The Expanding Force in Newton's Cosmos, as Shown in His Unpublished Papers* (Madrid: Ediciones de arte y bibliofilia, 1981), 15.
5. Castillejo, *The Expanding Force in Newton's Cosmos*, 78.
6. Betty Jo Dobbs, "Review of David Castillejo 'The Expanding Force in Newton's Cosmos, as Shown in His Unpublished Papers,'" *Isis* 73 (1982): 268.
7. Castillejo, *The Expanding Force in Newton's Cosmos*, back cover.
8. Westfall, *Never at Rest*, 239.
9. The phrase is from a letter from Newton to Nathanael Hawes, May 25, 1694, in H. Turnbull, ed., *The Correspondence of Isaac Newton* (Cambridge, UK: Cambridge University Press, 1961), vol. 3, 360.
10. Ivor Grattan-Guinness, "The Role of an Editor: Some Remarks on Whiteside's Edition of Newton's Mathematical Papers," *British Journal for the History of Science* 43 (2010): 105–12.
11. Grattan-Guinness, "The Role of an Editor," 108.
12. See Kenneth Knoespel, "Interpretive Strategies in 'Theologiae gentilis origins philosophiae,'" in *Newton and Religion: Context, Nature and Influence*, edited by J. Force and R. H. Popkin (Dordrecht: Kluwer, 1999), 193.

## Epilogue

1. Email to the author from Arnold Hunt, July 24, 2012.
2. See Harrison, *The Library of Isaac Newton.*
3. De Villamil, *Newton: The Man.*
4. Paul Quarrie, "The Scientific Library of the Earls of Macclesfield," *Notes and Records of the Royal Society* 60 (2006): 5–24, citation 6.
5. Paul Quarrie of Sotheby's, quoted in "Feud Forces Sale of 'Intellectual Time Capsule,'" *Telegraph*, February 20, 2004.
6. [Augustus De Morgan], "Review of *Correspondence of Scientific Men of the Seventeenth Century, in the Collection of the Earl of Macclesfield*," *Athenaeum*, October 18, 1862, 489–91, citation 489.
7. On the disunity of Newton's archive, see Rob Iliffe, "'A Connected System'? The Snare of a Beautiful Hand and the Unity of Newton's Archive," in *Archives of the Scientific Revolution*, edited by Michael Hunter (Woodbridge, UK: Boydell Press, 1998), 137–57.

# Index